A Year's Work in an Out-Apiary

An Illustrated Guide to Beekeeping Colonies and Collecting of Honey from Hives

By G. M. Doolittle

PANTIANOS
CLASSICS

Published by Pantianos Classics

ISBN-13: 978-1-78987-269-9

First published in 1908

Contents

*D*edication

This book is affectionately dedicated to
W. Z. HUTCHINSON,
the most strenuous advocate of keeping
more bees in out-apiaries, with
the least possible labor,
of the present time.

Preface

Immediately upon the publication of "Scientific Queen-rearing," In 1889, I began work upon the problem of non-swarming, either with or without manipulation, putting fully as many years and as much thought into this problem as I did to bring about the matter published in the queen-book. It is not my intention, in this work, to go over all the ground "traveled" during the past sixteen years, used in fully perfecting the plan as given in this book, as the book is written for the specialist, and particularly for one having or desiring to keep bees in out-apiaries.

While the book is intended for the specialist, it is none the less desirable for the plain, every-day bee-keeper, with his one home apiary, or for the amateur with his five to ten colonies; and because this book is for the specialist in bee-keeping I have not gone into first principles or the A B C of our pursuit, as the specialist has passed these rudimentary things long ago. There are plenty of good books before one, and all who are desirous of learning of the foundation structure, therefore, have no need of repeating here. (The amateur should certainly procure, read, and digest one or more of these books upon entering the ranks of apiculture.)

For these reasons I have "jumped right into the arena," without fear or asking any favors, and told the reader, in as simple language as I could, and as briefly as possible with a full understanding of the plan in sight, just what was done at the twelve different visits I made to the out-apiary, during one year, and the results accruing therefrom; and I do not think that I have exhausted the subject, but I have given the *first principles* the same as I did in "Scientific Queen-rearing." While I have been digging out the plans as given in this book, others have been building many different structures on the queen-rearing principle I gave, for which I am glad; but they have not undermined the principle, nor brought it to naught.

And now I send this non-swarming-section-honey-producing principle out, expecting that others will build different structures on it; and if they do, I shall be glad again. I cheerfully and freely give the principles in this system to all, hoping and believing that the same will prove as efficient in the hands of others as they have with me.

Chapter One - An Average of 114½ Pounds of Section Honey per Colony in a Poor Season, and How It Was Done

The sun rose bright and clear on the morning of April 14, 1905, the morning of my sixtieth birthday; and as old Sol peered over the hill-top in all his golden splendor, kissing the swelling buds and cheering all animated nature with the intuition that "spring has come," I proposed to Mr. Clark, my partner, that we go over to the out-apiary, five miles distant, and set the bees out of the cellar, the bees in the home apiary having been set out two or three days previously. The horse was soon hitched up, as the roads were too muddy and full of deep ruts for the auto, and we were at our destination before nine o'clock, with the stands all prepared for the bees.

Dr. Miller's bottom board, summer side up

As I use the Dr. Miller bottom board, the same having a two-inch-deep side for wintering, and a three-eighths-inch-deep side for summer, a reserve board was placed on the stand of No. 1, row No. 1, summer side up, for the first colony taken out to be placed upon. Before going to the cellar, two smokers were set to burning — one in the bee-yard, side of stand No. 1, row 1, and one at the cellar door just outside. Besides this last smoker, there was placed a soaking-wet (right-sized) piece of cotton cloth at the cellar door, ready for immediate use as soon as any hive was brought out, for there is nothing that will keep bees from pushing out of their hive before you want them to like a wet cloth.

Colony No. 1 was now brought through the cellar door; and while Mr. Clark shut the door, so the bees remaining in the cellar might be kept as quiet as possible, I put the wet cloth over the entrance of the hive, and then sent a few puffs of smoke in at the entrance through a little hole made by turning back one corner of the wet cloth. A loud roar soon told that this colony was in good condition. A rope was now put under the cleats to the hive, when the same was carried to the bee-yard (Mr. Clark on one side, Doolittle on the other), and set down by the side of the stand it was to occupy. The crate staples which held the bottom-board to the hive were now pulled out by a prying motion with a piece of a wagon-spring, fitted so it would just slip through between the wood of the hive and the top of the staple; then a little more smoke

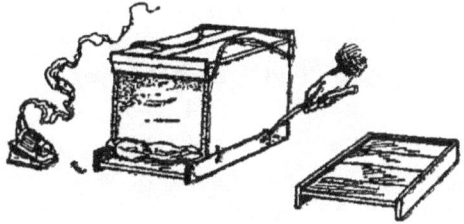

Removing the Staples

was used to drive the bees back so they would not be in too big haste to come out of the hive when the same was lifted from the bottom-board it had previously occupied, to the one on its own stand. The use of smoke in this way I consider of great advantage; for since so using I have had no mixing of bees on their first flight, no matter about wind, clouds, or how warm the day was; nor whether each hive was set on its old stand or not. I see by the bee-papers that others have much trouble with bees mixing when setting out, so that a part of the hives have colonies strong beyond measure, and other hives correspondingly weak, as used to be the case with me before adopting this smoking plan. In the home apiary almost any plan can be adopted; but when we go to an out-apiary a plan must be

Use of Rope in Carrying a Hive

used that will allow us to set out the bees on *that day* and at *that time* without danger from mixing so that a part become too strong and a part too weak, and the above is that plan. The bottom-board No. 1 had occupied all winter was now turned summer side up and placed on stand No. 2, when colony No. 2 was brought out in the same way No. 1 had been, and placed on it, and so on till all were out.

In this way all dead bees, dirt, etc., which had accumulated during the winter were at once done away with, leaving all sweet and clean, and in a prosperous condition. Owing to too much work of all kinds for two or three years previous when I was alone, the bees at the out-apiary had been allowed to become reduced in numbers to 21 colonies in the fall of 1904, one of which was queenless and weak in bees. An inventory taken after all were out gave 11 good colonies — 5 fair, 3 weak, and 2 dead — the dead ones being the queenless colony, and one which was made by setting an upper story off late in the season, thinking it had a queen because the bees did not go down through the bee-escape when it was put under to rid the hives of honey from bees.

Entrances Adjusted

After the bees had settled down a little from their first flight, two or three that seemed light in stores when putting them on their stands were fed by exchanging two frames of empty combs for two full combs of honey left over from the year before. The entrances were now adjusted to suit the size of the colonies, 3 inches by 3/8 being given the good colonies; 2 by 3/8 for the fair, and ¾ by 3/8 for the weak.

Next each hive was looked after to see that all was as tight as possible about the top, and that everything was in good condition for leaving till pollen became plentiful from the elm and soft maple, when we started for home. This was what was done on the first visit for the season of 1905.

Chapter Two

Ten days later, April 24, the elm and soft maples were in full bloom; and as the day was fine I went to the out-apiary again, arriving there about nine o'clock. I found the bees were almost rolling over each other, carrying in the yellowish-green pollen from the elm and the greenish-pink from the maples.

During the season of 1904 some 70 combs of honey in Langstroth frames, averaging about 7 pounds each, had been left for spring use, after seeing that all colonies had enough for winter, these being kept, with other combs more or less empty, for a purpose which will be explained further on. Each hive was now opened, beginning at No. 1 on the first row, to see that each had a good queen and honey enough to make them "rich" to a prosperous degree till fruit-trees came into bloom, from three to four weeks later. Any colony that did not have 20 pounds of honey was given one, two, or three of the seven-pound combs till it did have that amount; and if any colony had more, none of it was taken away, as there is nothing which gives better results in bees in the spring than to have the colony so rich in stores that it feels no need of retrenching.

Very many, at the present time, seem to think that brood-rearing can be made to forge ahead much faster by feeding the bees a teacupful of thin sweet every day than by any other method; but from many experiments along this line during the past thirty years I can only think this a mistaken idea, based on theory rather than on a practical solution of the matter by taking a certain number of colonies in the same apiary, feeding half of them while the other half are left *"rich"* in stores, as above, but without feeding, and then comparing "notes" regarding each half, thus determining which is the better to go into the honey harvest. And some go even further than this, claiming that it is a very paying operation to extract the honey from the brood-combs which are in the hive, and then thin this honey and feed it back again to the bees — reasoning that brood-rearing can not go on prosperously with combs of solid honey acting as "great cold barriers in the midst of the brood-nest," and also that "solid combs of capped honey in the middle of the brood-nest are surely in the way of a prosperous increase." I can not understand such reasoning as this when coming from men who stand high in authority — men who have or should have a thorough knowledge of the inside of the brood-chamber, and especially the inside of the *brood-nest;* for never during my nearly forty years of manipulation of brood-chambers did I ever once see *even one* "solid comb of capped honey" in the "middle of the brood-nest" or "in the midst of the brood-nest," during the spring or early summer, unless the same was inserted there by the hand of man. Bees do not allow such a state of affairs; and when the hand of man thrusts a solid comb of honey in the middle of the broodnest, the first thing the bees do is to uncap such honey and carry it into the cells outside and surrounding the brood, filling the inserted comb or cells occupied with honey only a day or two before with eggs and larvae so that there is an additional lot of brood in these cells. This fact led to what has been known as the "spreading of the brood," which has been both praised and condemned at intervals during the last half-century. I have tried the feeding and the spreading of the brood plans by the side of the "rich-in-honey plan," as is given in this book, setting apart a given number of colonies to each, only to prove, after numerous trials and experiments, that the latter would outstrip either of the others in the race for brood, as well as saving all feeders, all the labor of extracting and feeding, which amounts to a *great big* lot when the same is practiced on fifty, one hundred, or more colonies. Just put the combs of honey in next the sides of the hive, shoving the frames composing the brood-nest into the center, if it is not already there; and this once fixing of the brood-chamber is all the work necessary for the largest amount of brood the reigning queen can produce.

In fact, after trying all other plans for stimulating brood-rearing at the out-apiary I feel to say there is nothing that comes "anywhere near in sight" of this plan of "millions of honey at our house." All talk about daily feeding, as practiced by Mr. Alexander and others, or the spreading of brood, as I have advocated during the past, is of no use at the out-apiary, for the apiarist can not be there to attend to it. And, even if he could, results show that the "millions of honey at our house" plan, followed by what is to come hereafter, will outstrip any of the heretofore known stimulating plans by far in the race for bees in time for the harvest.

I have *dwelt* on this matter, as I consider it one of the most important things when an out-apiary is worked for comb honey. After seeing that all had 20 pounds or more of honey, the entrances were enlarged to about 5 Inches in length for the best colonies, to 1½ inches for the weaker ones, or enough so the bees would not be crowded till the next visit, which would be nearly a month later, or during fruit-bloom. A careful inspection of the brood showed that all the queens were good, as all brood was in compact form, with eggs on the outside of the outside combs to the brood-nest (not brood-chamber), all the cells within this circle of eggs being occupied with one egg in each.

I have noticed for years that a poor or failing queen does not lay like this, but "scatters" to a greater or less extent according to her poorness. "Where I find queens that are poor, as I sometimes do at this time of the year, they are killed, and one of the weak colonies, with its good queen, is united with the colony from which the queen has been killed. By attending to this queen matter when taking off the clover or basswood honey each year, superseding all queens more than three years old, and those younger which may show signs of failure, the problem of poor queens in the spring is practically solved. Far better supersede at that time than in the spring.

This is something well worth "pasting in the hat." A careful look over the yard, the last thing, to see that all was in "applepie" order, and I was soon gliding in the auto over the road toward home at the rate of fifteen miles an hour, which is fast enough, considering the roughness of the roads and our hilly country. The above was what was done at the second visit.

Chapter Three - Bloom Time

As I looked out over the valley, and to the hills beyond, on the morning of May 20, 1905, a beautiful sight met my gaze. The dew-covered grass, in many fields, was glistening in the morning sunshine, while the plum and cherry trees, with their white flowers, in the orchards, nestled down among the more showy apple, whose pinkish-white bloom so ladened the air with fragrance that, from sight and smell, one could hardly think but that he was in the sinless "Eden land" when the "stars sang for joy" on creation's early morn. But a neighing from the barn calls out "horse to be fed," and the "rounds of another day" are begun. After breakfast the horse was hitched up, as the roads to the out-apiary are too muddy, from the rain of the afternoon before, for comfort with the auto.

Arriving, I find the bees starting out in good earnest for the apple trees, which is just what I want, as they will now be out of the way when I am hunting for the queens, for to-day's work is to consist in part in finding and clipping the wings of all unclipped queens. This clipping part would be wholly unnecessary with the plan to be given were it not that, owing to certain peculiar seasons, the bees in a few colonies will take it "into their heads" to swarm a few days before I am ready to do the "swarming;" and in such cases as these, where all queens have their wings clipped, these colonies are held together until the time has fully arrived for making swarms. As such peculiar seasons do not come oftener than about one year in four, I have sometimes thought I would give up the clipping; but so far I have adhered to it, very much as a man will stick to the insuring of his buildings when there has not been a "fire" in his school district for forty years.

There are many ways of finding queens for clipping or otherwise; but after trying all I much prefer the following: Take a light box with you, the same size as the hive, only three inches wider, so as to allow plenty of room for the combs. After looking over the first comb, set it in the box, next to the farthest side of the box, always sitting or standing with the back to the sun, and having the box and hive so the sun will shine on the "face" side of the combs next to you. On taking out the second comb, quickly glance over the "face" side of the next comb in the hive, and if the queen is there she will be seen running to get around on the dark or opposite side of the comb, she being easily seen in the strong sunlight when thus moving. If not seen, immediately look on the opposite side of the comb you hold in your hands, when this comb is to be set next to the other in the box. Now lift the next frame, looking first on the "face" side of the next frame in the hive, and then on the opposite side of the frame in your hands, as the queen will be, in nineteen cases out of twenty, on these "dark" sides. In this way keep on tin the queen is found, or till all the combs are in the box. If all the combs are in the box before you find her, look the bees over that are in the hive; and if not found then, commence to set the combs back in the hive, looking as before at the two "dark" sides. I find forty-nine out of every fifty queens looked for, before the combs are all in the box, and the fiftieth one before they are all back in the hive again.

On opening the hives I find the honey quite largely turned into bees and brood, as only the two outside combs have much in them — six to eight combs in each hive being nearly solid with brood, except those which were weak in the spring. That the colonies having eight frames of brood need not contract the swarming fever before I visit the apiary again, and that all may be as nearly equal as possible when the bloom from white clover opens, I take one of the most nearly full frames from these — a frame composed of nearly or quite all sealed brood, from which I see a few bees just beginning to emerge — and put the same in one of the colonies having but six frames of brood, putting the nearest empty comb this colony has, taken to make room for this frame of emerging brood, in the colony from which the brood came. In this way all are made as nearly equal as possible. As brood-rearing has been going on now for about a month, the hives are so well filled with bees that there is no danger of any setback from a cold spell; and if we are to stop all swarming entirely except in the occasional season referred to above, no swarming being a thing most ardently desired for an out-apiary, if not an actual necessity, we must now "pave the way" for the same by commencing before the bees have any thought of the "swarming season."

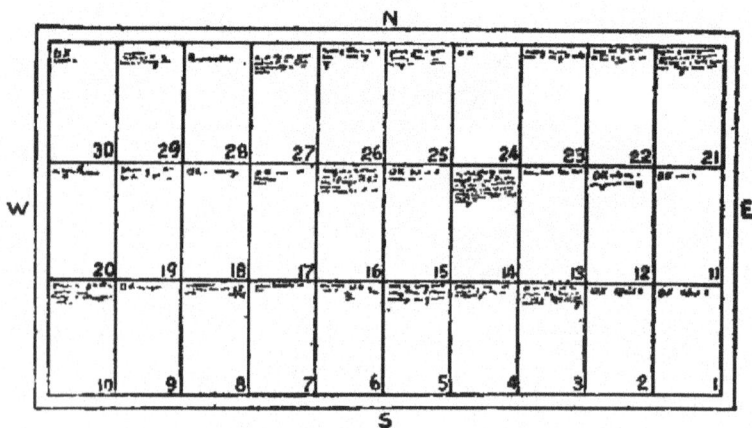

Doolittle's Record-Board for the Apiary

12

After clipping all the queens, and fixing the brood as above, and having jotted down on the 8x16x¼-inch smooth board I have carried with me the condition of each colony, I sit down a few minutes to outline the season's work from what the board shows. This board has on it, in miniature, a sketch of the whole out-apiary — each row of hives, and each hive in its place, shown in squares on either side. Each square is numbered the same as the hives, and in these squares I make a record at each visit, giving by brief signs the condition of each colony and its needs, slipping the board under the cushion to the seat of the vehicle I use in going to and from the apiary.

In this way I have the exact condition of the apiary spread out before me at any time I may wish to know about it. I now find that 13 of the 19 colonies have 7 combs of brood each, and are good enough to receive an extra story at this time; and these, together with three others, are set apart for section honey, or 16 in all; the three weak colonies (and nine others to be made later) are to carry out the other part of the plan, to be given later on.

How to Make Comb-Honey Colonies at an Out-Apiary "Rich" in Stores for Brood Rearing

So far I have been working for the largest possible amount of brood which will give bees in great numbers at the time of the honey harvest, and there must be no slackening now if success is to crown my efforts. To this end, and to keep the colonies from getting the swarming fever, I use a ten-frame Langstroth hive. Small hives, the hiving of swarms on a full set of startered frames, so they will not swarm out, and later taking half of them away, so as to "send" all the white honey into the sections through the con-

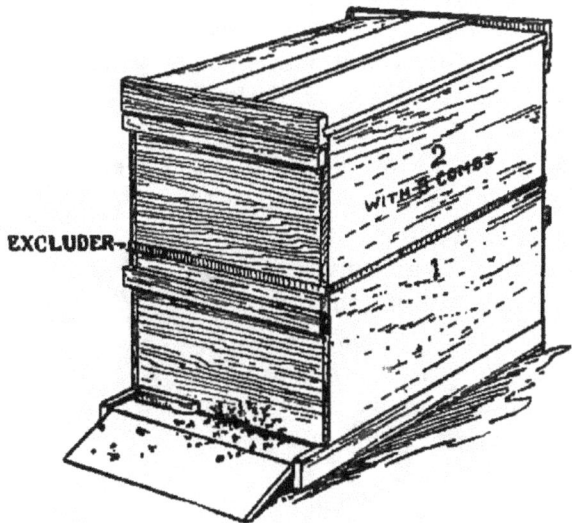

A Two-Story Colony "Rich" in Stores for Brood-Rearing

traction of brood-chambers; the turning of the parent colony one way and another every few days, after the prime swarm has been cast, so as to throw all the bees emerging therein with the swarm, etc., may do very well for the home apiary; but any thing which requires so much manipulation, watching, and care has no place in a non-swarming out-apiary. In fact, with the plan I used to produce 114½ lbs. of section honey per colony in 1905, about the poorest of all seasons in this locality during the last 30 years (acknowledged by the editor of *Gleanings* to be the shortest crop in the United States in many years), the ten-frame hive is to be preferred to any thing smaller.

Nearly all that has been written during the past was from the "viewpoint" of the home apiary, under the swarming system. W. Z. Hutchinson has well said "that few of the writers in the journals write from the point of view of the extensive bee-keeper — the man with out-apiaries. So many times I remark to myself when reading the description of a method, 'That's all right when a man is in the apiary all the time, but it won't work in an out-apiary.'" Just so. I have found while working out the plan as here given that very nearly all of my writings during the past were of no practical importance when working an out-apiary on the non-swarming principle, with a view to the greatest possible amount of comb honey with the best possible labor. But, to return:

Having decided that 13 colonies are now ready for "treatment", I go to No. 1 and take out the two outside frames, containing mostly honey and pollen, putting two empty combs from the reserve pile in their place. I now put on a queen-excluder, and on top of this I set another ten-frame hive, having eight combs in it, the same being more or less filled with honey, just in accord with the way these reserve combs come oft the colonies the fall previous. Perhaps I'd best tell right here how I get these reserve or extra combs. Wired frames were filled with foundation and given to colonies to draw out into combs, till I had an extra set of ten combs, or twenty nice worker combs for each colony I expected to work at the out-apiary for section honey, each year. To return again.

Having the hive with eight combs in it, set over colony No. 1, prepared as given, I take the two combs of honey taken out, and shake the bees from them so as to be sure the queen is not gotten above, when two of the eight combs in the upper hive are placed a bee-space apart, toward one side of the hive, when one of the combs of honey is put in. Four more of the eight combs are now drawn toward the frame of honey just put in, properly spacing them, when the other comb of honey is put in, the other two combs spaced, and the hive closed. The diagram shows the arrangement.

I now fix the other twelve colonies in the same way, when all are ready to do the best work possible in every way till white clover blooms. Taking the years as they average, and fixing each upper hive with an average of the reserve combs, as to honey for each colony, each will have from 15 to 30 pounds, and this amount together with the way their "riches" are fixed, and the bees straightening up things to their liking, gives a zest to brood-rearing which soon very nearly or completely fills the ten combs below, and that in time to give the maximum amount of bees in the clover and basswood flow. If the bees do not secure honey to any amount from the fruit-bloom, mustard, or locust, on account of bad weather, as is often the case in this locality, they go right on with their brood just the same, as the amount of honey they have demands no retrenching. Then these combs act as a sort of balance-wheel; that is, if a short flow of nectar comes for a day or two, there are empty cells in abundance in which to store it; and the bees do not hesitate to take all that is needed for the most prolific brood-rearing, if the next two days or a week are days of storm, cold, or an entire failure of nectar. Thus we have no days of crowding out of the brood with a sudden flow of nectar on one hand or a slackening or failure of brood on account of "famine" on the other hand; while at the same time this doubling of the hive room entirely prevents any of the colonies contracting the swarming fever before the time for the working-out of our plan just when it will give us the assurance of the most perfect success.

The Importance of Getting the Colonies in the Spring in the Best Possible Condition for the Harvest

Again, I wish to quote from W. Z. Hutchinson: "Can you bring your bees through the spring and have them in the best possible condition for the harvest when it comes? Are you sure there is nothing you can do in this period to increase your crop? I came across a bee-keeper a short time ago who secured a crop far in advance of his neighbors; and the only difference in his management, so far as I could discover, was that he fed his bees between fruit-bloom and clover; and when the latter came the combs were full of brood and food, and the surplus went into the supers at *once*; besides, there were more bees to gather it." This is just what this plan, as here given, accomplishes. The bees are abundantly fed, so there is no slack in brood-rearing; the combs in the lower hive (ten in number) are full of brood. There are nearly double the bees to gather honey when the harvest comes that there are when working by the old plans; and about the honey going into the supers *at once* — I will let the worked-out plan tell you further on.

If a good yield happens to be obtained from fruit-bloom, wild mustard, and black locust, the brood-nest or lower hive is not crowded with honey, as would have been the case had not this upper hive of combs been given, for the combs of honey raised from below and put above tell the bees from the start, "This is our storehouse," and there is room enough in it to hold from 60 to 75 pounds of surplus, above what was in the hive when I closed it. With a good flow from fruit-bloom or any other source, just at this time, together with the honey that we had allowed them at our former visit, had they been kept in the lower hive, with no sections put on, would come a material lessening of our prospect of a surplus from clover and basswood, either from forcing them to swarm prematurely or the crowding of the queen, by filling the cells with honey, which should be occupied with brood. Elisha Gallup was right when he told us, years ago, that such would be the case where a large surplus was obtained early in the season, from robbing or any other source, which filled the combs with honey before they were fully occupied with brood.

As now fixed, brood-rearing goes on "swimmingly," with no desire for swarming, and this is just what is desirable at any out-apiary (or home yard also) worked for comb honey. The entrances to all hives but the weaker ones are "thrown wide open," while these are given as large an entrance as the stronger ones had at the last visit before this. The "door-yard" boards are fixed so that the grass will not "swamp" the hives or hinder the bees' flight before' my next visit, and I am off for home. The work part, as given here, is what I did at the third visit.

Chapter Four - How to Control Swarms When Running for Comb Honey

It is now the 16th of June, many heads on the white clover are fully in bloom, while the black locust, from which the bees obtained quite a little honey, has just gone, and two of my bee-keeping neighbors report "swarming commenced." Half past three o'clock a.m. finds me in my auto, with the scythe done up in a blanket (to keep it from cutting and marring something it was not intended for), occupying the "other seat." Those who have never ridden in an auto at the "peep o' day" can not even imagine my delight that morning. Birds were singing from every branch, the barnyard fowls were out after the "early worm," while now and then the smoke from the chimney of an early, enterprising farmer was rising up in wavy circles as it ascended toward heaven. The eastern sky soon became all aglow with its "gold and carmine," telling of the advancing sun, while the cattle and sheep on a "thousand hills" were securing their morning repast from the grass made so pleasant and palatable from the "dew of the morning." When nearing the apiary, a jolly, fun-loving farmer, who had "just pulled out" for his cows, accosted me, while pointing at the wrapped-up scythe, with, "Taking the sick one in your ambulance to the hospital?" "Yes," I replied, without stopping, as every moment was precious, if I was to get the bee-yard mown before the bees got "waked up" by the rising sun.

Arriving, the scythe was hastily unwrapped, and, going to the front or south side of row No. 1 (hives face south, rows run east and west), I begin at the east end and mow a swath toward the west, allowing the "pointing in" to come as near each hive as is possible without hitting any of them. By thus mowing, the swath of grass is carried out and away from the entrances to the hives so the bees need not be disturbed when I come to the raking-up part, later on. Arriving at the west side of the yard, I quickly go back to the east or "beginning" end, and mow through again. Immediately in front of row No. 2, but instead of coming back "empty," as before, I mow back at the rear of the hives on row No. 1, cutting as close to the backs of the hive as is possible without hitting them enough to disturb the bees materially. In this way the double swath of grass is left in the center, between the first and second rows of hives.

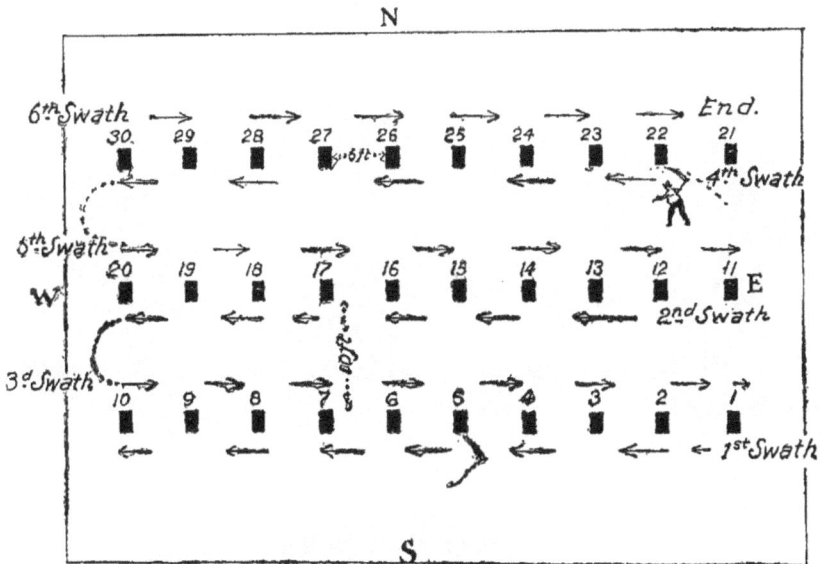

Doolittle's Method of Mowing the Grass in a Bee-Yard

17

I now begin in front of row No. 3, coming back at the rear of No. 2. Next, I go to the west end of No. 3 and mow at its rear, turning the swath away from the rear of the hives against the fence at the rear of the apiary, the same as I turned the first swath away from the entrances or fronts of the hives, against the fence in front of the apiary. I now go to row No. 1 and cut the grass between hive No. 1 and hive No. 2, and so on till the grass is cut between all the hives in each row.

After years of practice and experimenting, this is the best and quickest way to cut the grass in any bee-yard laid out in rows, that I know of; and, after a little practice, very little grass will be left about any hive to cut with a knife, shears, or sickle. Upon reading the above when published as a serial in *Gleanings,* Mr. W. O. Perkins, Bristol, Conn., writes: "I should like Mr. Doolittle's reason for going back 'empty' after mowing the first swath in front of row No. 1; also after fifth in rear of row No. 2. If after mowing in front of row No. 1 he will commence at the west end of row No. 1 and mow a swath in the rear of the same row, and come back in front of row No. 2, then back from west to east in rear of row No. 2, and back to west in front of row No. 3, and return in rear of

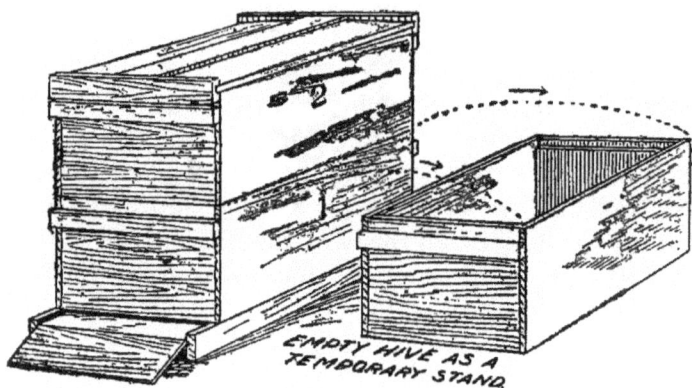

Use of Hive as Temporary Stand; 2 About To Be Transferred Thereto

same, he will save two trips across the yard 'empty' and therefore do the mowing in less time than by his plan." I wish to thank Mr. P. for calling attention to this matter, as his plan would seem the quicker to many; and it would be not only the quicker, but the *better,* if room enough could be afforded so the rows of hives could be 14 to 16 feet apart. But as 10 feet between rows is all that can well be afforded, considering both land rent and travel, other things come in that hinder mowing as Mr. Perkins proposes. A man naturally cuts a swath about seven or eight feet wide, which is little less than the distance between hives when the rows are ten feet from center to center; and if we cut back of row No. 1, or any other row, as Mr. P. proposes, the swath of grass would be thrown on and into the entrances of the hives on row No. 2, resulting in our seeing which was the "quickest" and "best" way out of the apiary, rather than the quickest and best way to cut the grass or to save time. By the plan I have given, the swath of grass is thrown against the back side of the hive first, and consequently the bees take no offense thereat. At half past six the grass is cut, raked up, and put in a pile outside the bee-yard fence, for the farmer who owns the land to use, if he so desires, and I am ready for my breakfast lunch, which I eat sitting In the auto.

That the reader may better understand, I will say that the fence enclosure is calculated for 30 colonies, three rows with ten on each row. The rows are ten feet apart from center to center, and the hives are six feet apart in the row, which distance I prefer to any thing else, after having tried distances both less and greater. When I bought this out-apiary it had only 22 colonies in it, and as I thought at that time that I did not care to increase the number to more than 30 colonies, it was laid out and planned for that number. And as, later on, I was

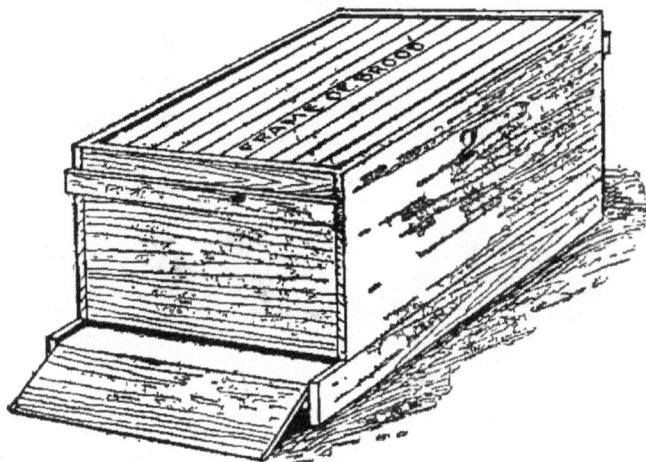

Center Frame is Taken Out and a Frame of Brood Put in its Place

overworked to an extent that retrenchment was considered rather than enlarging, it has remained the same as when first laid out. My first object when buying this apiary was, the forming of nuclei for queen-rearing at the home yard, as bees, no matter what their age, brought four or five miles from home, do not return so as nearly or quite to spoil a nucleus newly made, as do the bees taken from the same apiary.

My experience, based upon the time taken to work this 30-colony apiary by the plan here given, is that from 60 to 75 colonies would be the right number for each out-apiary to be worked by one energetic man, in a fairly good locality, without any help from others, except in setting in and out from the cellar. One man *can* do this, but I consider it money well spent when paid for help to do this carrying part.

I always begin any work with bees, where I can work in rotation, at hive No. 1, row 1, for this reason: if any colony becomes unduly disturbed at any time during any manipulation or work about it, I am soon behind

19

and away from their range of flight, so am less liable to be tormented by angry bees; for if the object of their anger is out of sight of the entrance of their hive, they soon forsake the following of that object. Any colony after being worked at, or after having work done about its hive, is much more liable to resent having a moving object in front of them, and in line with their flight, than they are before being disturbed. Here is also "another something" which is well worth "pasting in the hat."

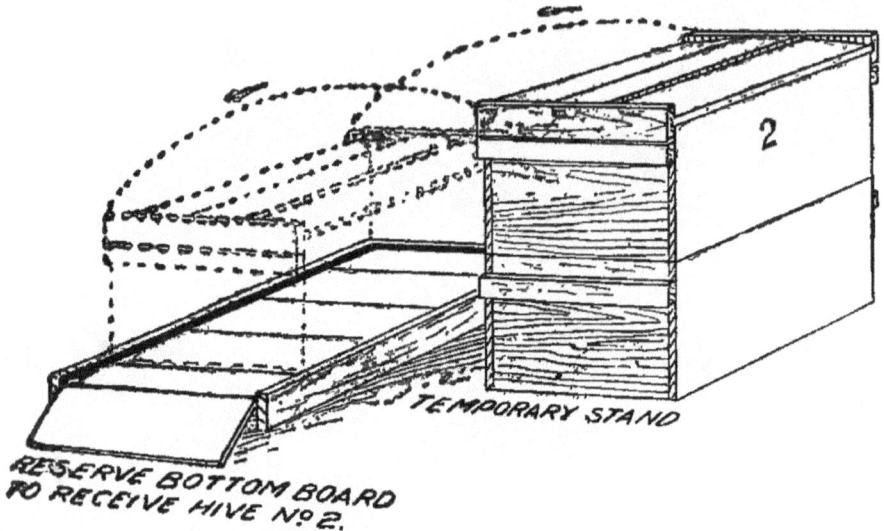

Lunch over, I take an empty hive and go to hive No. 1, this being one of the 13 having an upper story put on during the third visit. The empty hive is put down on the ground, close by the side of the colony, as a temporary stand, so that by a lifting, swinging motion, the upper story can be easily set on It with scarcely bending the back, *(See cut on page 18.)* which is now done, after prying it up at the back side and sending a few puffs of smoke under it to quiet the bees. From the strength it takes to "swing" this upper hive, I judge there is some 50 or more pounds of honey In It, which is more than I expected from the poor season we have had so far. The lower hive, bottom-board and all, is now set off the stand, and a reserved bottom-board placed thereon, when the upper story of 50 pounds or more of honey is set on this new bottom-board and a center frame of honey taken out, which is taken to one of the weaker colonies and exchanged for a frame one-fourth to one-third full of brood, with the rest of the cells nearly empty, after the bees have been brushed from each into their respective hives. This one-fourth-full frame of brood is now set in the hive on the new bottom-board, to take the place of the frame of honey. *(See cut on page 19.)*

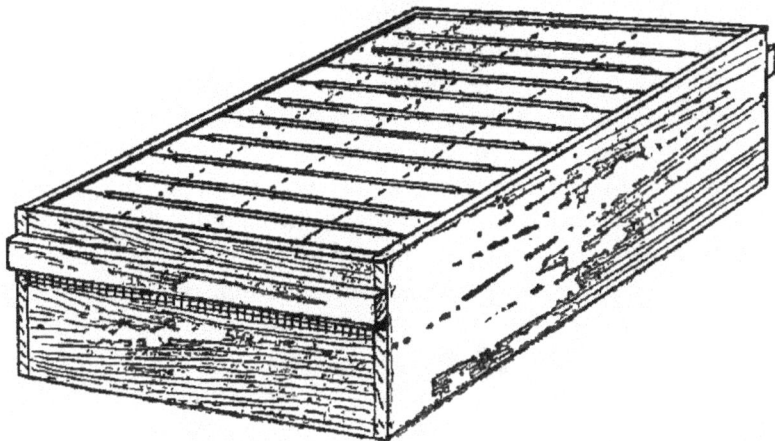

Doolittle Super Containing Forty-Four 3¼ x 5 3/8 x 1 5/8 Sections.

Such a comb seems to be quite necessary, where the upper hive contains much honey, as it establishes the brood-nest in the center of the hive, where it should be, and also allows the queen to keep right on laying without interruption, the same as she has been doing. If the queen is checked in her laying at this point, as she would be in a hive thus filled with honey, if no frame having empty cells was given, it is quite apt to result in an effort being made at swarming, which is not consistent with the immediate moving of the honey in these combs to the supers above, and the success we wish to obtain, although even in case of such swarming, better results are obtained than by any other plan of "shook" swarming which I have tried; for after a fruitless effort or two (the queen having her wings clipped so she cannot go with the swarm), and a few days of sulking, they will go to work with a will, thus showing their acceptance of the situation. However, if treated as here given, not one colony in fifty will do aught but accept the situation, and go to work at once in the sections, especially if there is any honey coming in from the fields, and the colony did not contract the swarming fever before it was shaken. From 35 years of close watching I find swarming to be conducted in this way as a rule: Queen-cells are formed, or the walls of old queen cups drawn out till quite thin, when the queen lays eggs in them. In three days these eggs hatch into larvae, which are fed a little less than six days, when the cell is sealed over. On the day after the sealing of the first cell the prime swarm issues with the old queen. In order that the queen may fly and accompany the swarm on the wing when the first larvae are about three days old, she begins gradually to cease laying, and almost or entirely stops three days later, or at the time the colony would naturally swarm. When the swarm finds a home, and comb-building begins, the queen slowly begins to lay.

Increasing every day, till at the end of four days she has arrived at her usual prolificness again. And thus the queen is usually from seven to ten days in a partial or wholly resting period, as to egg-laying, at the time of swarming, the same being entirely necessary where increase is done without the interference of man. Now, through sickness at the time of "shook swarming," in 1906, which made me leave a part of the colonies till they contracted the swarming fever and swarmed, I learned that, through the stopping of the queen from laying for swarming, before a colony was shaken, and the consequent four days before she arrived at full prolificness again put that colony in the same condition as the one which contracted the swarming fever after shaking, because no frame partly full of brood was given. For this reason I have emphasized, further on, the shaking of all colonies that are strong enough at the commencement of the honey-flow, before they contract the swarming fever. Of course, the upper hive of combs retards preparation for swarming for a long time; but if the shaking is not done till the bees begin to crowd honey into the combs of the brood-chamber, thus restricting the laying of the queen, swarming is the result, the same thwarting to a greater or less extent the full success of the plan.

I now get two supers of sections, from the pile which has been brought, 8 to 12 supers at a time, each time I have come to the apiary, either with the horse or auto, each super containing 44 one-pound sections, as this is the number of 3¼ x 5 3/8 x 1 5/8 sections my super covering a ten-frame Langstroth hive contains. The sections in one of these supers contain only full sheets of foundation, this foundation being of the extra thin kind for sections, as manufactured by The A. I. Root Co., while the other super has 12 of the 44, full or nearly so of comb, left over from 1904, as "unfinished" sections, they having been put in the super when it was prepared for the season of 1905, with the 32 other sections (filled with extra thin foundation) as "baits."

The baits are very valuable with this plan, as these bait-combs give a chance for the bees to be storing the honey at once, or immediately after its removal from the combs below, to give the queen room for her eggs — this being done while other bees are drawing out the foundation in other sections, so that work along all lines progresses as one great whole, without any interruption.

This super with baits is put on first, top of this prepared hive, and the other with sections of foundation on top of this. This second or upper super is put on to give plenty of room for any overflow of bees or honey which may come before our next visit, so that the bees may not at any time feel crowded for room. I have sometimes put the super containing the baits at the top, but the bees do not so readily get to them there, and hence slower work all around on the start, which is against the greatest success. An immediate start in the sections is a great advantage at this stage of proceedings.

Some seem to think that the bees carry very little if any honey from the brood-nest up into the sections, and for this reason feed the bees. between apple and clover bloom, sugar syrup or inferior honey till the combs and all cells not occupied with brood are solid full of sealed honey, claiming, for this plan, yields in excess of others who do not so feed, arguing that, by such a plan, all of the clover honey is put into the sections.

Doolittle's Wide Frame with Sections

I believe they make a mistake. The greater yield comes from this fed honey or sugar going up into the sections, together with that gathered at this time from the clover bloom; for I have repeatedly seen such combs of sealed stores all emptied, or nearly so, two weeks after the clover began to yield honey, eggs, larvae, and sealed brood taking its place. And another thing which proves to me that very much if not all of the honey which is now in this prepared hive goes into the sections, is the color of that stored therein during the time the queen is

Doolittle's Queen-Excluder with Drone-Hole

filling the hive below with brood. Quite a little of the honey left over from the year before, which is given to the bees for stimulative purposes, so that they may think there are "millions of honey at our house," was gathered from the buckwheat. Of course, the most of this goes into brood; but often there is enough left, so that, when mixed with the other honey which has been accumulating since these combs were put in the hives during the spring, together with that coming in from clover at this time, it will give the honey in the sections from the first super filled a very delicate pink hue, and a taste not quite like clover in its purity. I thought at first that this part would be against my worked-out plan, but I find that these pinkish-colored sections (the same being detected only by holding them up and looking through them toward the light) are preferred to any of those which are filled with only clover or basswood honey, and coming off later, they often selling for a cent or two more a pound than the white, to those knowing about them.

Having every thing now in readiness, the hive is closed by putting on the cover, when the queen-excluder is taken off the hive of brood, and I at once proceed to shake and brush the bees off their combs of brood in front of this prepared hive.

Perhaps I better say a few words regarding this shaking off part, for very many do not seem to be handy at shaking the bees off their combs. some of our best bee-keepers telling me that it could not be done to any advantage till I had shown them how. Let me see if I can tell the reader so he can do it with ease. Let the projecting ends of the top-bar to the frame

24

rest mainly on the big finger of each hand; then, with a quick upward motion, toss these ends against the ball of the hands at the base of the thumb, and at just the instant the ends of the frame strike the ball of the hands give the hands a quick downward motion. This takes the bee off its guard, as it is holding on to keep from falling off the comb downward, having no idea that there is any danger from falling off upward. But this "falling upward" is exactly what it does, as three-fourths the bees, when I shake the combs, are tossed up in the air as they are dislodged. The instant the ends of the frame strike the fingers again, toss it up against the ball a second time, and then back to the fingers, when, if you get the "hang" of the matter, as you will after a few trials, you will find that 990 out of every 1,000 bees are off the comb; and if you have that proportion off you will have no need of the brush, for it Is not necessary to get each and every bee off the combs of brood. Only ten to fifteen bees left on each comb will be

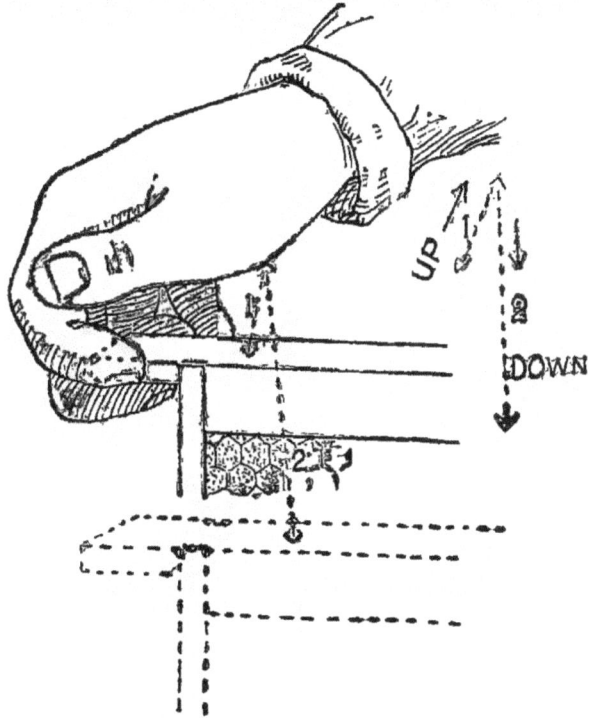

Doolittle's Method of Holding and Shaking Frames

but from 100 to 150 bees for the whole, which will make little difference with the swarm. However, I like to get as many as is consistent with quick work, with the shook colony, for the more bees there are here the better results in honey. Then, I wish to say that there are times when thin nectar is coming in bountifully, when I cannot shake all the bees off thus, or by any other plan; for if I do the bees will be so nearly drowned in this thin nectar which shaken out of the combs that they will not go in the hive. During such a flow of nectar I shake the combs the same way, only do it so gently that no nectar is shaken out, when the bees which still hold to the combs must be brushed off. I have always declared it a nuisance to have thin nectar coming in at a time when I am obliged to free combs from bees, but have always been consoled by the thought that this thin nectar Is what is to be turned into cash by and by, when the bees have it evaporated into nice honey, so enjoyable to the consumer later on. Where I am obliged to use a brush I greatly prefer the "Dixie," as sold by The A. I. Root Company, to any other I have ever tried, and I have used all which have been advertised, and many besides, which have been sent me for my approval and recommendation. This brush is soft, so that it does not injure the bees, and yet is firm enough to take all the bees off one side of the comb with only one stroke over the same. In the absence of any brush, through oversight or something of the kind, a bunch of five or six tops of goldenrod, or even of grass, will do very well — in fact, better than some of the brushes which have been sent me.

As the hive into which this "shook" colony is to go is really their own home, and contains more than an abundance of honey, this plan does entirely away with all the labor and time used in drumming and pounding the hives, as well as waiting for the bees to fill themselves with honey — something which has been considered as a thing of vital importance with all of the other plans of "shook swarming." Nothing of the kind is required to make the swarm stay, or for any other purpose, for the bees are still on

Shaking the Bees in Front of the Hive

their own combs, with sufficient brood, and room enough for the queen to lay right along. All of this, together with the carrying of the honey from the combs into the sections, keeps them contented, and brings great results in honey to their keeper. As a prolific queen in the height of her egg-laying always falls off the comb she is on at the first shake, she being so heavy with eggs, I hold each comb as low as possible in front of the entrance in shaking, so she shall not be injured by the fall. I have reason to believe that many queens have been seriously injured by the "shook swarming" of the past, through the carelessness of the operator In this matter, and then the plan condemned, because, at the end of the season, more honey has been found in the brood-chamber than in the sections, when the operator alone was to blame for the queen being injured to such an extent that she could not keep the combs filled with brood, as she otherwise would. Always remember that a good queen is the chief source of success in all things pertaining to a large yield of section honey. As fast as the combs are freed from bees they are set in the empty hive, at first brought, each comb being set in the order it had in the old hive till all are in.

I now go to one of the colonies considered too weak to tier on the third visit, take off the cover, put on the queen-excluder just taken off from No. 1, and on top of this I set the hive of beeless brood, when the cover is put on top of all.

If, in shaking the bees off their combs, I come across any comb which is not more than one-third to one-half full of brood, the same having as many empty cells as those containing the brood, such a frame is kept out of the hive of beeless brood, and used to go in No. 2 when it is prepared the same as No. 1 has been, instead of a comb taken from a weaker colony, as was done when fixing No. 1. As this saves time, can be found in very many of the colonies, and answers the purpose just as well, I am constantly on the lookout for such during the time I am "swarming" the colonies.

Returning to No. 1 the bees that still adhere to the empty hive and bottom-board are dislodged, so as to fall with the rest of the "shook" colony, when the hive and bottom-board are carried to No. 2, which is to go through the same process as has No. 1, and so on till the whole 13 have all been "swarmed," which takes far less time for each one than the telling how it is done. By this plan I do not have to look for the queen nor overhaul the combs, nor by any other plan look for queen-cells, as is generally the case with most of the other ways of artificial swarming. In all the other plans of "shook" swarming it is recommended to wait about the "swarming" till queen-cells are sealed, or have eggs or larvae in them. In my practice I have found that this is all a myth, and it is also something that is not applicable to the work in an out-apiary, with only a few visits to the same each year. Yea, further: I find it an absolute detriment to have the bees prepare to swarm before this "shook swarming" Is done. About the middle of June, 1906, I was taken sick, as before spoken of, so that only about half of the "swarming" was done at the time it should have been, the rest of the colonies being left till nearly July before I was able to finish. During this time more than half of those not "swarmed" contracted the swarming fever, built queen-cells, and swarmed; but as the queen's wings were clipped they could not get away. This was the condition I found them in when I was able to get to the out-apiary to finish the swarming that year. Now for the result: The queen in such colonies that had swarmed, or gotten queen-cells sealed, or nearly ready for sealing, had nearly or quite stopped laying, as queens always do at the time of natural swarming; and when "shook" they did not go to laying prolifically *at once*, as do those shaken just at the commencement of the white-honey harvest; consequently such colonies kept the queens restricted as to egg-laying, after she commenced to lay later on, by not carrying the honey into the sections, and their showing in section honey was less than two-thirds of that given by those which had not prepared to swarm, and little more than half of that given by those made before I was taken sick.

All that is necessary is to have all the colonies, to be treated, strong to overflowing with bees. Then, when the time is ripe to do the work, *go on and shake*, no matter about the queen-cells, whether they have them or not; only, if any are found with eggs, larvae, or pupae, in them, when the combs are shaken and freed from bees, they should be cut off, so they will not bother by emerging in the hives above the queen-excluder. Nor can the idea that the colony that starts no queen-cells, and would not swarm if let alone (the same giving better results if left unshook), be tolerated or carried out in an out-apiary when worked for comb honey on the "few visits" plan, even if this "giving better results" was the case, as the liability of such colonies swarming at unexpected times must always be counted upon. But such Is not the case when the apiary is worked on the plan here given, for nearly all of the colonies; treated in this way give better results than any colony which does not swarm, worked in the usual way. There-fore this way of working, as here given, does away with all the labor re-quired in trying to find out whether colonies are going to swarm or not, by way of looking for queen-cells once a week in using the different plans that have been published, such as tipping up the hives and smoking the bees so the bottom of the combs may be inspected for cells, clamps for holding the sections from falling off while this inspection is going on, the prying-apart of divisible brood-chambers to see if queen-cells are being built between, or even having a "cell-detector hole" cut and fixed in the back of the hive, which can be opened once a day or oftener to discover if cells are started on a comb, cut and fixed in such a way that, if queen-cells are started in any part of the hive, they will be started so they can be seen from this hole; and, also, all the labor of requeening, caging queens, etc., used in trying to prevent swarming. In fact, it supersedes any and all the plans heretofore used by hundreds and thousands of apiarists when working on the shook-swarming plans or prevention of swarming. And as it not only does away with all but a minimum amount of work, and also gives the greatest possible yield of section honey, I claim that what is here given stands "head and shoulders" above any thing else in sight dur-ing the past or at the present time, especially in working an out-apiary for comb honey.

Hive No. 3, on row 2, contained what I considered my best breeder for comb honey, and I had left it strong in bees and brood on the last visit, hoping it might make some preparation for swarming by the time I came again, and when shaking it I found just what I wanted, which was queen-cells with one or two day old larvae in them, the larvae literally swimming in royal jelly.

The shaking of these combs was done more carefully than with the others, for fear of dislodging the swimming larvae, although there is little danger along this line, until the royal larvae attain an age of four or five days. This hive of beeless brood and queen-cells was placed on top of the strongest colony not tiered on my third visit, and only this one hive of brood was put on it, while the others had two and three hives each, as I had 13 hives of brood to go on six colonies, hence taking only one for this hive left twelve to go on the other five not tiered before. Why I put no other brood on this colony with these cells was because I wished these royal larvae given every possible advantage looking toward the best of queens. I do not generally depend on queens or queen-cells from the out-apiary for work therein, as I generally have more time and conveniences for rearing them in the home yard, taking whatever I wish along this line with me at each visit. But if we have a good breeding queen at the out-apiary, and wish to use cells or queens from her brood, as was the case above, this tells the reader how it can be done. By the way, here is an excellent plan for the amateur to raise queens for use in his own apiary: When a colony having the *best* breeding-queen is found preparing to swarm by having queen-cells with from one to four day-old larvae in them, just take the frames having these cells on them and place them in an upper story over a strong colony having a queen-excluder under the upper hive, and see how nicely they will complete them. When the queens are ready to emerge, set one of the frames having a "ripe" cell on it, together with a frame of honey, bees and all, into an empty hive, and in ten or twelve days you will have a good nucleus with a fine laying queen; and by treating all in the same way you can have as many nuclei as you had cells. Then by giving frames of brood from other colonies to take the place of those taken from the best breeder she will again have more queen-cells with larva in them in a week or so, when these in turn can be put on the strong colony over the excluder till the cells are ripe, when more nuclei are made, and so on till we have all the *"best of queens"* we need.

The reader is undoubtedly familiar with the truth advocated of late years, that, if an extracting-super is placed over a colony as soon as it becomes strong in bees, swarming will be retarded to quite an extent. Then on the arrival of the honey harvest, if this extracting-super is taken off, and a super of sections placed on the hive, the bees will the more readily enter the sections from the fact that they have been used to working above the brood-nest. I practiced this quite largely eight to twelve years ago, and obtained much better results than I had done before. Ever since "Scientific Queen-rearing" was given to the public (1889) I have been spending my best efforts in trying to work out a perfect plan of non-swarming, either with or without manipulation; and during the first six or seven years, just as I would begin to think I had something of value a different season would come, the bees swarm, and spoil it all. I was about to give up in despair, when one day it came to me, "why not use this extracted super plan, combined with shook swarming?" which was then first appearing in sight. My mental reply was, "I do not want any plan that will not put the first-gathered honey (more than is needed for brood-rearing) anywhere else than in the sections." Then the thought came, "Is it not possible to have the first honey, which others extract, stored in the upper story of a full-sized hive, thereby retarding swarming still more, and then work in such a way as to cause the bees to put it in sections later on?" With this, despair turned to hope, and this hope has become a reality by the perfect working of the plan as now given to the public; and the result of the year 1905 (114½ lbs. of section honey on an average per colony), the poorest of all late years for honey in this locality, has caused me to write the matter up, so all who wish can use it.

Having the 13 colonies "swarmed," and the six others on the road to prosperity after a careful looking-over the whole, to see that "all is well," the scythe is again wrapped up, allowed a whole seat in the ambulance (auto), the starting-crank turned, when I am soon experiencing a delightful rest in the "noonday" sun (which had seemed pretty hot in my work in the bee-yard), made so comfortable through the pleasant breeze caused by the tireless running of the automobile. In this we have what was done at the fourth visit although the same is pretty well mixed up with other things pertaining to the developing of this plan.

Chapter Five - A Simple and Reliable Plan for Making Increase

Just ten days have elapsed since I started on my fourth visit to the out-apiary, and I am getting ready to go again; but this time I am obliged to go with the horse, on account of its raining seven days out of the ten. So much rain has caused the roads to become almost impassable on account of the mud; and the almost constant rain at this time has caused the bright prospects of an abundant harvest of honey from white clover, which has been more plentiful than usual, to fade nearly out of sight. It does not rain this morning; but it is cool and cloudy, with a fine mist in the air. Such a day is not adapted to working with the bees to the best advantage; but it is necessary to go today, if I am to save those nice queen-cells, which are of much advantage to me just at this time. After a steady splash, splash, splash of the horse's feet in the mud for nearly an hour (as I cannot drive "off a walk"), we arrive at the apiary.

Having put the horse in the farmer's barn I now proceed to place nine of the reserved bottom-boards, and as many covers on as many unoccupied stands, when I go to the hive having the brood from the best breeding queen that had the queen-cells with the little larva in them at the last visit, and, upon examination, I very luckily find that six of the ten combs have one or more fine, nearly ripe, cells on them. From one frame having four cells on, and two others having five, I cut two cells from each, and "graft" them into three of the frames having none, putting the frames back in place again. The clouds are now "breaking" in the sky, with the sun peering occasionally through the mist, which tells me I am to have a fairly good day for my work, after all — far better than I had even hoped for. I now take one of the frames having queen-cells on it, together with the bees on the same, and carry it to one of the hives having the tiered-up brood, taking from this a frame (bees and all) and putting the one with the cells in its place. In all this work with tiered-up brood, when changing the same from one hive to another I do not disturb the bees on them, as bees above a queen-excluder are, to all intents and purposes, queenless, so make no trouble by putting them in different hives. It is best generally to put the frame having queen-cells on it near the center of the hive, as this seems to give the better results.

Having the frame with queen-cells in the hive, I next take the frame of brood and go back to the hive having the cells, when it is put in the place left vacant there. In this way I keep on until the five colonies having upper stories of brood have a frame with queen-cells on it from the best breeder.

I now take off these five prepared upper stories, setting each on one of the bottom-boards previously placed where they are to stand, putting on the covers and adjusting the entrance to about three inches in length. The setting-off of these hives paves the way for using the other four frames having queen-cells on them in four more hives of brood, following the same plan in treating them which was used with the five now fixed on new stands, for the making of that many new colonies, so I have nine more colonies than I did when I entered the apiary an hour or so before. As the brood in these combs is all sealed now, and the bees on them are nearly all young bees, with more emerging every minute, there will be no setback to this colony from the bees returning to the colony they came from, as is generally the case with the most of the ways used in making colonies by the "setoff" plan. And this is the best, quickest, and easiest way of making colonies with which I am familiar; and this I say after using it for more than ten years, and after having tried nearly all the plans given by others.

Entrance Contracted to Three Inches

If for any reason I wish a greater number of colonies than can be made as here given, and wish them for the purpose of taking care of bee-less brood, I make as many as I think I shall need, during my third visit to the apiary. In the following manner: I take two frames of emerging brood from the colonies having eight frames, and, instead of giving them to the colonies having the six combs of brood, as I told about in giving an account of that visit, I put them in a hive, after having brushed the bees off, together with two or three of the reserved combs - one, at least, of which should contain honey. The space left vacant where the brood was taken from, in the strong colony, is filled with two combs from the reserve pile, thus giving the queen in this colony room for more eggs. I now go to another of the stronger colonies and put a queen-excluder on it for the time being, when this prepared hive, having the two combs of emerging brood, is set thereon, where it is allowed to remain two or three hours, during which time the young bees come up from below sufficient to care for the combs and brood, after which it is placed on the stand I wish it to occupy. When I expect to make colonies in this way, if I have no laying queens thus early in the home yard I send south for them, if it is possible to get them from there. A queen-cell *will answer,* but the laying queen is much better.

32

By the way, full colonies can be made in this way at almost any time of the year when there is plenty of emerging brood by taking two combs of such brood from three or four strong colonies and adding to these, frames of honey. I have made such with perfect success as late as September first, using six combs of brood and four of honey. It is so easy — no hunting of queens nor any thing of the kind; and the best part of the whole is, enough of the young bees *always* stay to make it a success. No need of natural swarming for increase when we can make as many colonies as we desire in such a simple, easy way. The advent of the queen-excluder was a great blessing, and one of the needed helps in giving us the "modern apiculture" we now enjoy.

Then there is another way of making new colonies about the time of the early flow, or ten days before it commences, just in accord with the strength of the colonies; when we want an increase as well as all the section honey we can secure, as is often the case when a number of out-apiaries are to be built out of the first one started. Let us suppose that it is from five to ten days before the expected flow is to arrive, and that, in accord with this plan, no upper stories have been put on. The bees have built up from the extra honey we have allowed them at the outside of the hive until they are in good shape, and the swarming season is drawing on. We now go to hive No. 1 and take out the comb of brood on which we find the queen, bees and all, unless this comb contains mostly maturing brood. If we find her on such a frame of maturing brood, the frames are looked over until we find one which is only partly filled with eggs or larvae, when the queen is put thereon; and this frame with the bees and queen is set aside till we are ready for it. In this way we know right where the queen is, so we can work rapidly without danger of losing her whereabouts or injuring her. All of the rest of the brood is left in the hive where it is; and if the combs are not all occupied with brood, those not so occupied are taken out, and frames of brood are taken from some of the colonies which may be too weak to work in sections during the season to advantage, and put in this hive, until it is filled with combs, every one of which has brood In it. "We now put on top of this hive of brood, still on its old stand, a queen-excluder, and on this a second hive or upper story, as such a hive is generally termed, when the frame of bees, brood, and queen which we set to one side at the beginning, is placed in the center of this upper story, the same being filled out with frames of comb, a few of which should contain some little honey, after which the hive is closed. Brood-rearing will now leap ahead in this upper hive from the heat, and bees which will come up from below to feed the queen and encourage her on, while the bees cannot swarm, as the queen cannot pass the excluder, so there is no need of any worrying on our part for ten days. In from eight to eleven days, just as the weather will permit (as no young queen will be likely to emerge from any queen-cells the bees may rear from the brood in the lower hive previous to the eleventh day), we visit the out-apiary again, at which time we set the upper story containing the queen, bees, and new brood off to a new stand which we wish a colony to occupy. This setting off causes all the field bees to return to the old stand, which makes it a powerful colony. It will be noted that the hive on the old stand now contains nearly all sealed brood, with no larvae young enough to turn into a queen, even should the eighth day be the one on which we wish to go to the out-apiary; and only sealed brood, should the weather be such that we do not go until the ninth, tenth, or eleventh day.

Having arrived at the hive, after setting off the upper story it is opened and all queen-cells destroyed, if any are found, and the colony given a ripe queen-cell from the best breeder; or a virgin queen can be introduced, if preferred, as this colony is hopelessly queenless, after the cells are cut. A super of sections (with baits) should now be put on, and on top of this another super of sections. The bees will not swarm, as they have but the ONE queen-cell, or virgin queen, and there are no eggs or larvae from which to rear another. I have not tested the matter; but I am quite sure that a laying queen would be just as good, if introduced by the candy-cork plan at this time, in a locality where the honey harvest is not of such long duration as to cause the bees to swarm with her later on; for this colony is practically in the same condition as a colony ten days after swarming, except much stronger. With such colonies, when using natural swarming, I have often given laying queens in this way, and always had better success in comb honey than I did where all the cells were cut but one. This giving a laying queen at this time would also do away with the only trouble which attaches to the plan when a ripe cell or virgin queen is used, which trouble arises from the fact that occasionally some queen will fail to get back from her mating-trip, which may not be found out in time to save the colony with the few visits we make at the out-apiary. As soon as the young queen begins to lay, the honey will go up into the sections with a rush, as the emerging brood reinforces the field laborers for ten to twelve days after the old queen and her hive of new brood was removed to its new stand. The removed colony will usually become strong enough to store sufficient honey for wintering from the basswood bloom, and often a surplus of twenty-five to sixty sections is obtained from buckwheat, should the season prove favorable for the secretion of nectar from this source. In the same way as we have treated No. 1, as many others are used as we wish increase of colonies, and in this way we not only secure the needed colonies for other out-apiaries, but a good crop of section honey from our bees.

Hive filled with Reserve
Combs and placed over
a colony.
Queen excluder between.

Colony.

With the making of the nine colonies, as above given, I have the desired number for the year 1905, as I have house-building and other work going on, so I have no desire for further enlarging this year. I still have four colonies with a hive of brood on each, the one having completed the queen-cells being the stronger. As I wish to work sixteen colonies for section honey, and having shaken only thirteen at my last visit, I now prepare to shake three more. To do this I pick from the reserved combs enough to fill three hives, using those the nearest full of honey. One of these hives is now carried to the colony completing the queen-cells, a reserve bottom-board placed on its stand, after it has been set off, and the hive with combs of honey set thereon.

A comb only partly full of brood is now selected from the upper story, one from which many young bees have emerged, and more rapidly gnawing from the cells, this being set in the center of the combs of honey; then two supers are set on in the way those were at the fourth visit, when I proceed to shake and brush the bees off from the whole of the nineteen combs still remaining in the two hives; then from the hives and the bottom-boards, thus giving this colony all the bees from two hives of brood, or, as a rule, very many more than those had that were made at the fourth

35

visit. After two more of the strongest colonies have been treated in the same way the beeless brood is tiered up on those remaining, when a moment of taking an inventory shows that I now have sixteen "shook" colonies, two others containing three hives of brood and one of four hives, the queens of which are confined to the lower hive by the queen-excluder, and nine colonies just made, having queen-cells ready to hatch, together with nine frames of brood, which will all emerge in eleven days, making twenty-eight colonies in all. In order that the remainder of the reserved combs may not be destroyed by moths they are now placed, ten in a hive, and one set on top of each of the twelve hives not having sections on them, a queen-excluder having first been placed over the nine just made colonies not having any on. The year 1905 was an exceptional one, in that the colonies in the apiary had been allowed to become so few through overwork.

When the whole thirty, fifty, or seventy-five stands (or whatever number we decide upon for an out-apiary) are occupied at the time of setting out in the spring, there is no need of making colonies as here given. When we have the full number, four-fifths of the best colonies are worked for section honey, while the weaker one-fifth are to care for the beeless brood, and combs, which become the "reserve combs" in the fall, for the next season. That the reader may understand more fully, suppose that the out-apiary is laid out for seventy-five colonies, and that we have that number in the spring; then we shall want sixty hives of reserve combs to go on to the four-fifths of the stronger colonies, which in this case would be sixty, the work with each being done as given in chapters three and four.

In thus working, these sixty hives of beeless brood will be stacked on the one-fifth, or fifteen colonies, where they will remain till the end of the honey season, when they are taken off and stacked away for reserve combs for the next year, as will be given later on. This will make each of the fifteen colonies have five hives of brood, the queen being confined to the lower hive by the queen-excluder. At first glance it would seem that some of this brood would be neglected through the giving of so much to one colony; but repeated examinations prove that all is well cared for. As the weather is warm at this time of the year, and as many young bees are emerging from these combs every hour, a few bees on the start can hold things in perfect condition till all danger is past. When this brood has all emerged, such hives have an army of bees, which, in a good season, often fill all the hives with honey, thus giving us an insurance for the next year when that needed for brood is so used, and the rest of it carried up to the sections, so there is no loss. It will be noticed that, by this plan, *all the honey* not used in the *actual production* of bees goes into the sections (which is something no other plan heretofore given ever accomplished), that the bees and queen are stimulated to their utmost In early spring by this large amount of honey telling them *"millions of honey at our house,"* so that there is not only no loss by having these combs stored full after the brood emerges, but a positive advantage through the stimulating effect they have the next spring. Mr. E. R. Root, editor of *Gleanings,* writes me that he fears the plan as given In this book may be a failure In many localities on account of there being no buckwheat-honey flow in the fall to fill these combs for spring use; but I think his fears are groundless, and I guess he has forgotten the grounds taken in the past, both by himself and his father, "that very strong colonies will store a surplus of honey where weak or only fairly good colonies will hardly make a living." These four and five stories of brood turn out a colony of *enormous proportions* during the next two months, with an *"army"* of bees marching in and out at the entrance while ordinary colonies are doing little if anything; and the result has always been plenty of honey in these combs for use in spring, or *"millions of honey at our house,"* even after the *poorest* season for fall honey I have ever known; while in good seasons from 200 to 300 pounds to each colony has been the result, each of the four upper ten-frame Langstroth hives being about all a man could lift when piling up in the fall. If all of the sixty colonies were not ready for treatment on my fourth visit, then I put one or two hives of beeless brood on top of those not quite strong enough in bees to shake, setting this brood under the hive of reserve combs they have, so the brood will all be together. This gives such a colony so much extra room that they will not think of swarming during our next ten days' absence, notwithstanding the vast numbers of bees emerging from these two or three hives of brood.

Late Shook Swarms for Comb Honey

When I go to make the fifth visit the reserve combs are set down on the bottom-board, and the bees from all three hives are shaken out. This gives rousing "shook" colonies; and if a heavy yield of honey is on just at this time, these later-made colonies will even surpass those shaken at the fourth visit. In section-honey production; and it sometimes happens that the yield of honey will make It profitable to shake colonies having three and four stories of brood, right at the beginning of the basswood flow, thus bringing nearly or quite 100,000 bees in one of these hives of reserve combs, quite well filled with honey, in which case three and four supers of sections are used to give the proper amount of room for their working to the best advantage. However, this requires an extra visit, which may not be convenient when we are working a long string of out-apiaries.

A Hive Prepared on the Doolittle Plan

After having tried this way two or three times I often think it is just as profitable to let the honey go into the reserve combs. But the section honey stored by such a rousing colony, right in the height of basswood bloom, is so perfect and handsome in appearance that my mouth often "waters" for such, and the eagerness of consumers for the same makes it very profitable for market. When it is thought desirable to use this late plan of shaking, colonies can be formed by the plans given, which will care for the brood, and if desired they can be wintered over to take the place of any that may die during the winter. Then if none die they can be united with others, so that the number may be kept at the thirty, fifty, or seventy-five, decided upon when the yard was laid out.

Occasionally there will come an extremely bad season for the bees, like that of 1907, when it kept cold and rainy nearly every day up to the first of June, and it seemed almost impossible to have any colonies strong enough to take advantage of the white-clover bloom when it came. Of course, in such seasons the bloom will be somewhat delayed; but as most of our grasses thrive quite well under cool wet weather, this bloom is not delayed to nearly the extent that the brood in the hives will be; therefore, if we are to secure any section honey at all during these very bad years we must work on the retrenching plan rather than on one of expansion. At the time for putting on the upper stories of combs partly filled with honey in the season of 1907 there was not a single hive that had brood in eight combs, while the majority had brood in only six combs, and some of those not nearly filled. In looking over the situation I see that the only plan that gave me any assurance of success was that of *"massing the brood,"* as I call it, which was done as follows:

When the cover was off the hive ready for putting on the queen-excluder I took out all of the combs which did not have brood in them, putting these in what was to be the upper hive, leaving out as many of the reserve or wintered-over combs as was necessary to make room for these. In those hives in which I found brood in only six combs, the other four combs were put in the upper story, or the hive which was to be soon put over the queen-excluder. Having these thus arranged I went to another hive having brood in six combs and took out four frames which were the nearest occupied with emerging brood of any they had, and put them in the colony from which I had just taken the four combs having no brood in them, while the four reserve combs which had been left out of the upper story were put in the now weaker colony to take the place of the brood taken out. These four combs were placed each side of the two combs left having brood in them, rather than in the center between them, so that the queen in laying, which she will now do rapidly, will work out from these two frames of brood, and in this way no brood will be lost during such cool or cold weather, as would be the case were these four combs placed in the center, or from where we took the brood. By working thus, this now weaker colony will get into the right condition to receive a hive of beeless brood when the time of shaking arrives, as will soon be given. A queen-excluder is now placed on top of the hive we have just filled with brood, and the hive full of combs, four of which have just been taken from below, is now set over the excluder, when they are left till our next visit at the opening of the clover bloom. Hives which have seven combs of brood have their three broodless combs taken away and three frames of brood from another colony given them, and so on, until all which the yard will furnish, that are strong enough, are fixed in the same way. In this way these colonies which are given brood arrive in nearly as good condition for shaking, when the season for so doing arrives, as do those which are of our best in a good year, so that they are ready to take advantage of the clover and basswood bloom when the same arrives, while *NONE* would be did we not have an "eye" to the existing state of affairs. Of course, we can not secure as large a crop of honey from the *part* which can be gotten ready in this way as we could were all strong; but we *can* secure quite a crop of section honey, even in the poorest season, as against none, or very nearly that, did we not have an *eye to the season.* Then, to carry out this plan so as to make the colonies from which we took the brood do their part in the matter, at time of shaking, the combs of beeless brood are given to these colonies, *one hive to each,* in such a way that we do not lose as much as we would naturally think, while gaining a whole lot by getting the others strong in time to take advantage of the first honey-flow of the season. Lest any may think that this way of working *is the rule* rather than the exception, allow me to say that, nine years out of ten, we do not need to resort to anything of this kind, for we are more often met with the conditions of the colonies building up in *advance* of the season, when we are taxed the other way to keep down the swarming instinct till the time of shaking arrives. At the time of shaking the strong colonies a queen-excluder is placed over the weaker colonies from which we took the brood at our visit the last time we were at the out-apiary before; and after the bees are shaken off the brood from the strong colonies, their beeless brood is set on top of the excluders, where It remains for from four to six days, in accord with the weather, or our necessary work at some other place, when the out-apiary is visited again and a super of sections having baits is placed on top of the excluder, and on top of this the hive of brood we have removed to give room for the super of sections, which is now virtually between the brood in both hives, this giving the colony so much room that it does not think of swarming, even with the great increase the maturing brood in both hives gives in the number of bees. To overcome the difficulty of stained sections and cappings to the combs in the sections, a sheet of enameled cloth, or a sheet of tin (the latter being preferred), is placed so it will cover their tops and all the openings between the sections at the top, except those in the last row next to the front of the super, and here there is only a place of about two inches wide left which is not covered. Through this space the bees pass up and down to and from the brood in the upper hive, so that the brood is kept in as good shape as it was before putting on the super, and thus the colony soon becomes very populous, without desire to swarm, while this small passageway to the upper hive so nearly excludes it from the bees that work is soon begun in the sections, although more or less honey will be stored in these combs above as the brood emerges. But as this is to be used for turning into bees the next season, no loss occurs on this (storing in brood-combs) account. If the season gives prospect that more than the one super will be filled, another is placed above it, as in the usual plan, raising the tin covering the sections to top of this super, and the only objection which comes from this way of working in a very poor season is that this upper hive must be raised or set off when we wish to manipulate the sections in any way. In this way I secured during the extremely poor season of 1907 a little over sixty-one pounds per colony, on an average, while multitudes of colonies in this locality, which were allowed to take their own course, did not give a pound of surplus. If success is to crown our efforts under all circumstances, conditions of different seasons, and especially during an uncommonly backward spring, we must "have more than one string to our bow," so as to get the bees and the season together, even in the worst seasons which ever come; and by thus doing, a very poor season may chronicle success on the right side of the ledger page. And because I believe the above is the right "string" to pull for success in a very poor season, I have given this exceptional plan of working; while the other and more fully described plan is the one to be used unless the season gives promise of being very poor.

Why an Empty Super of Sections Should Be Put on Top of Rather Than *under* Sections Partly Filled

It is now nearly noon, with the sun shining brightly, and the air becoming warm and balmy. To see the army of bees rushing in and out of the hives containing the "shook" colonies is a sight to gladden the heart of any bee-keeper; and those returning from the fields seem quite heavily loaded, though the nectar Is very thin on account of so much rain. I tried to count those coming in loaded during one minute, but they dropped down so fast in almost bunches of threes, fives, and sometimes ten or more, that It was impossible to do it. I counted two hundred, and estimated that fully twice that number went in without counting. Such colonies as these will do something at securing nectar, even if it does rain the larger part of the time.

I now take a little time to look at the supers of sections, and a glance at them shows the honey being sealed in the bait sections, with the most of the other sections in the lower super, having the foundation fully drawn out and the honey sparkling in every cell, nearly ready for sealing. With all but two colonies the bees are well at work in the upper super also, drawing out the foundation, with now and then a section having quite a little honey in it. Those that are as far advanced as this have their supers exchanged — that is, the upper super is set directly on top of the broodchamber; and the lower one, having the baits, now nearing completion, is placed on top, after which a super of sections, filled with the extra-light foundation, is placed on top of the whole, so that in no case shall any colony lack for room.

In all of my working with the bees I have not found that the placing of an empty super over one in which the bees are at work is any detriment, as the bees seem capable of clustering in the openings at the tops of the sections they are at work in, thereby forming a crust of bees that holds the heat in the super they are at work in, to such an extent that the work goes right along.

This is done on the same principle that a colony in early spring is able to maintain a temperature of 93 to 98 degrees inside of the brood-nest (which is the proper temperature for brood-rearing), when the temperature of the hive all around the crust bees does not rise above 45 to 50, when we have a spell of freezing weather. A colony of bees seems to be capable of holding almost any degree of temperature it desires, simply through a crust of bees which often does not at any point touch the hive. How this is done I do not know. But I *do* know that a handful of bees, less than 1,000 by count, kept the temperature where their brood was, between two combs, at 93 degrees, when the mercury outside stood at from 18 to 26 degrees above zero during a cold spell in April. And I have known (many times when experimenting) of good work being done in the sections, fixed as above, when it was so cool that not a bee would be seen anywhere from or in the upper super, except the crust between the tops of the sections In the super below.

Since these experiments I have always kept these reserve supers on top, ready to catch any overflow of bees or honey. But the placing of such a super under one in which the bees are at work often proves a great damage, especially in a poor season. Therefore, as a rule, during late years I never raise a partly full super up from the brood-chamber unless I can place one underneath it, in which the bees have commenced to work more or less.

Those colonies which have not yet commenced work in the upper super, or have only just begun, are left as they are, as such have all the room they will need until the next visit. In changing these supers I can not resist the temptation to look into the brood-chambers of two or three of the colonies, and in doing so I find the comb given them as a "starter," which was from one-eighth to one-fourth full of brood when placed in the center of the hive at time of "shook swarming," ten days ago, is literally filled with brood, two-thirds of which is sealed over, while six of the remaining nine frames, which were nearly full of honey at that time, have three-fourths of the honey removed from them, while the emptied cells are teeming with brood from the egg to larva in all stages of growth. This shows that the colonies are in a very prosperous condition; and should favorable weather come, a good harvest of white honey may yet be obtained. After a careful looking over to see that all things are in good shape for leaving I say goodbye to the pets at the close of this, my first visit, to the out-apiary; and in the above the reader has a record of what was done at this visit.

Chapter Six - How to Save Unnecessary Lifting in Taking off Filled Supers of Honey

But favorable weather did not come and continue; for on the very next day in the afternoon another rainstorm commenced and bad weather continued the most of the time during the next eight days, at the end of which the clover bloom is nearly past. We now have a few days of fine bee weather, still and clear, with hot days and nights, which the bees improve as best they can on the few nectar-giving flowers which are still in bloom. The first blossom-buds on the basswood-trees commenced to open on the sixth day of July, and I hoped that the good weather would continue right along; but with the afternoon of the seventh a two-days' rain commenced, which kept the bees in the hive nearly all the time. It is now the tenth day of July, and fifteen days since my last visit to the out-apiary. As there is a prospect of a fine day I start to make my sixth visit to that enchanting place. Before going, however, I catch and cage three just-laying queens, from as many nuclei in the home yard, that I may be prepared to give them to any of the nine colonies I made at the last visit, which may, by any means, have failed to get a laying queen from the cells then given, taking them, together with a load of supers, with me. As the basswood is now nearing full bloom I am hoping for better weather, the same as the farmers are, who, all along the road, are opening out their hay, which "got caught" out in the rain. Arriving, I find the bees rushing out of and into the hives, almost like mad in their wild scramble for the basswood nectar, which, to me, seems so thin that it is hardly worth the gathering, owing to the bloom having been kept wet continually for the past sixty hours. While the "scramble" for this thin basswood nectar is just as great as was that for clover nectar at my last visit, yet the number of bees going into and out of the entrances to the hives has lessened somewhat, owing to the death by old age of quite a number of bees which were on the stage of action at the time the colonies were "swarmed," while, as yet, none of the emerging bees are quite old enough to become field workers.

The first work is to look after these thirteen colonies, so that, should there be any supers ready to come off, they can be put on escape-boards the first thing, this giving the bees time to leave the sections so these filled supers can be carried home with me. I find that each one of the thirteen has one super fully completed, ready to take off; and several of them have a second super nearly so; but as I wish to take oft no sections not fully sealed over, at this time in the white-honey harvest, these nearly filled supers are allowed to remain on the hive. The taking-off at this time is done thus:

Doolittle's Scheme of Using a Wheelbarrow to Save Heavy Lifting

I put on the wheelbarrow (every apiary should have a wheelbarrow ready for use at a moment's notice) an empty hive, and beside it I put an escape-board, and on this escape-board a super of sections filled with foundation. (see illustration, above) The wheelbarrow is now brought up close to one of these colonies that has a super ready to come off, when the supers which are not ready are set on top of the super on the wheelbarrow, and the completed super set on the empty hive. By using the wheelbarrow, and working in this way, there is little if any bending of the back when lifting the filled and nearly filled supers, so the work is done quite easily— in fact, with as little fatigue as Is possible, and very much less than will occur when supers, hives, etc., are handled from the ground. The supers being now all off the hive and on the wheelbarrow, they are rearranged in putting back as follows:

The one that was at the top, the same being the one which was put on at the last visit, if the bees have worked In it at all, as they have in nearly all of them. Is set back directly on the brood-chamber, and on top of this is put the one which Is nearly completed, and on top of the two I place the empty super, or super of empty sections, just brought on the wheelbarrow. The board having the bee-escape in it is now put on, and on this the completed super is set. Having things arranged thus, and working in this way, no useless motions are made or lifting done that counts for naught.

The cover is now put on, and another escape-board and super of empty sections gotten, when I go to the next hive, treating that in the same way if it is in the same condition. If I find that any have done less work, then the nearest completed of the two supers, not as yet fully finished, is set on the brood-chamber, the one little worked in top of that, and the empty one from the wheelbarrow on top of this, with the escape-board and completed super above the three, as before. The thing sought after is to give room in such a manner that we shall not have a lot of unfinished sections should the season prove poor from now on, and at the same time provide plenty of room for the largest yield from basswood that is likely to occur in our locality; or, in the terms of an ancient parlance, have the "pot" right side up, should there be a great "downpour of porridge." To show how completely this plan works, I will say that, in the fall of 1905, I had only eight unfinished or unsalable sections out of every one hundred that the bees made a start on; while during 1906 the bees completed so nearly all that there were less than four out of every one hundred completed, this number not being sufficient to provide the necessary "baits" tor the season of 1907. This is very different from what it used to be when I worked on the old tiering-up plan, when I would often have from eighty to one hundred partly filled sections to every one hundred that were completed. The old saying is, that "a burnt child dreads the fire;" and having been severely burned several times during the past by putting an empty super under a partly filled one, just at this stage in the basswood bloom, which resulted, through a poor season afterward, in my having all the sections in both supers worked in, yet none completed in either, I am, perhaps, over-cautious now on this point. However, I think it better to use great caution at all times about putting an empty super under a partially full one, and especially so after having found that by putting the empty one on top better results can usually be obtained. I next look after the three colonies made by "shook" swarming at the first visit, exchanging supers and adding the third, where needed, the same as was given when telling how the thirteen were treated at that time. These have the supers containing the bait sections nearly completed, and I am tempted to take them off, but finally conclude to leave them, which proved the best thing to do, the way the season turned.

Doolittle's Candy Method of Introducing

I now look after the nine colonies made at the last visit, and an examination shows that all have laying queens but one, so I have two of the three brought, to carry back home. To the one having lost its queen a frame of young brood is given, taken from one of the others, and one of the three queens I have in cages is given to it. The removable stopper in this cage is one which I call a "candy cork," which is made by boring a five-sixteenths hole through a piece of an old broomhandle one inch long, or some other piece of wood that will fit into a round wire-cloth queen-cage, the cage being made by rolling a piece of wire cloth, fourteen or sixteen mesh to the inch, around said broomhandle or the finger, and locking the edges so it will retain the size wanted. This "cork," made from the broom handle, has the hole filled with "queen-cage candy," made by stirring and kneading powdered sugar and honey together till a stiff dough is formed, as is described in all of our late literature on bee-keeping. This inch in length of hole is filled with the queen candy, so the bees can liberate the queen at a time when she will be likely to be accepted by the bees, and at the same time not require my coming to the out-apiary to look after the matter, as would be necessary by other ways of introduction. As a rule it will take the bees about twenty-four hours to eat the candy out of an inch in length of the five-sixteenths-inch hole, which is about the right length of time in this case to warrant safe introduction.

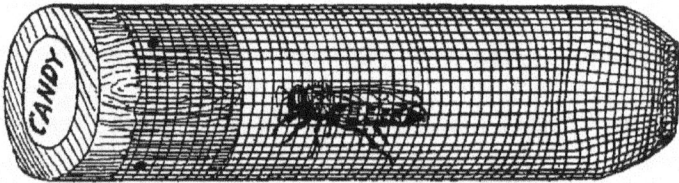

Doolittle's Home-Made Introducing-Cage

Having the queen all ready for the hive, a center frame is removed, and, after shaking the bees off, the cage is crowded between the bottom of the comb and the bottom-bar of the frame, seeing that there is nothing in the way of the bees having free access to the candy end of the cage, when the frame as thus prepared is set back in the hive and the hive closed.

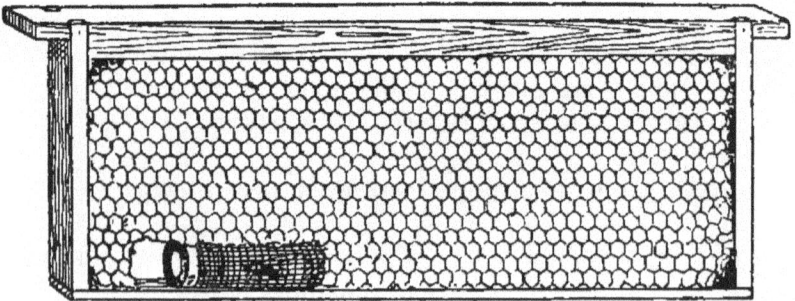

Manner of Placing Cage on the Comb

Owing to the length of time between visits, the above, and the giving of queen-cells, is about the only way that queens can be successfully supplied to queenless colonies at out-apiaries. If I think any of these new colonies, or those having upper stories of brood, will be apt to need more room than they have, I now put on at the top a hive containing wired frames filled with foundation, so that they can draw them out suitable for more reserve combs, and fill them with honey, should an extra good yield follow. In this way all are prepared for whatever may come, be the same wet or dry, cold or hot, a rich or a poor season, without feeling that I must go to the out-apiary with any change of weather that may occur.

What It Costs the Bee-Keeper to Let Grass and Weeds Tangle up the Entrance

I next look after the "dooryards" in front of all the colonies, making sure that none will be bothered in their flight by grass or weeds, as well as to look after any little odds and ends that may need my attention before leaving. This keeping of grass and weeds down in front of the hives is quite an item here in New York, as they spring up almost by magic in a wet season like this one. From some experiments I have made, by allowing the grass to "block" some hives for this purpose I find that, where badly tangled, the colonies in such hives will not store more than two-thirds as much honey during a good basswood yield as will those having a free flightway.

Dr. C. C. Miller, one of the closest observers in the bee-keeping ranks, thinks that a one-third loss is too much to attribute to a "tangled" entrance. But I cannot help thinking he has never seen hives as "badly tangled" up with grass as I have. The engraving does not do the "tangled hive" justice. Just imagine the grass growing a foot above the top of that

47

hive and super, and that so thick that you can hardly see through it. Then imagine how it would look after a south wind and rain had "lodged" it right over on the hive and entrance, and left it matted down there for from one to three feet over the "doorway" of the hive, and you will have only a fair idea of what I have seen in apiaries here in York State. Yes, I have seen this "tangle" so bad that it was fairly filled with pollen-pellets torn from the "baskets" of the pollen-gathering bees. And the strangest thing of it all was that the owners of these "tangled hives" called themselves "bee-keepers."

I am led to speak of this for the reason that I have found in many apiaries which I have visited during basswood bloom the bees crawling or hopping from spear to spear of grass or weeds, for from one to three feet from their entrance, before they could arrive at home with their loads. Heavy-loaded bees "tangle" much worse than those with no loads, and it seems cruel to make the little fellows struggle so to reach home, to say nothing about the apiarist's loss in honey. The looking after all of these things is often what makes the difference between success and a partial or entire failure. By this time the bees in the supers above the escape-boards will have nearly all run out of them, and the few remaining will go out during my wheeling them to the wagon, loading and getting started. The load and the mud make slow driving the order this time, and it is about 1:30 p.m. when I arrive at home.

In the above the reader has an account of what was done at the sixth visit. To be sure, there is considerable sermonizing mixed in, but this seems necessary for a full understanding of the matter.

Chapter Seven - Taking off The Surplus; What to do with the Unfinished Sections; Preparation for the Buckwheat Flow

It is now July 24, and the basswood bloom is all gone. With the exception of one or two days at a time, it has been rainy, cold, or windy all through its bloom. If possible the weather has been worse for the bees than during clover-bloom. If we could have had the good hot weather which came between the blooms, either in clover or basswood, a far different showing in honey would have been the result. Now that the basswood bloom is past, it is coming good weather again. While this can make no difference with me now, yet I am very glad to have it come, as it is cheering to the hearts of the farmers who have had an uphill time in securing their hay and winter wheat, much hay spoiling on account of the continued wet. Again I am off on the road to the apiary, carrying with me another supply of supers, for the buckwheat bloom is still ahead. As I go, my heart is made light through seeing the many fields on the hillsides and valleys covered with their waving grain, basking in the sunlight, while the pearly streams, being nearly at full bank from our recent rains, make sweet music in their joyous journey toward the river. The pasture lands are nearly as green as in June, while, generally, at this time of the year they are brown and bare. The farmhouses nestle among the green branches of the trees, giving prospect of garnered fruit through the half-grown apples, plums, and pears, discernible among the sun-kissed leaves. Surely all nature is happy — why not I? I have done my best with the bees; and if a meager crop is the result, through no fault of mine, I should be happy with what I get.

With such scenes and thoughts as these, the time passes almost too soon; and before I am hardly aware of it the horse is turning in at the farmer's roadway leading toward the bee-yard. With the horse stabled before a manger of rain-cured hay I enter the apiary. Each colony having sections on is looked after, fixing them now so they are supposed to be all right till the end of the buckwheat harvest, which is the end of the sur-plus-honey season in this locality. The wheelbarrow having i an empty hive, bee-escape, and super of foundation-filled sections, is again brought into use, when all the fully completed supers are set on the empty hive, and the others on the empty super, the same as with my last visit. If a su-per is found having two-thirds or more of its sections completed I think it best to take off the same, as those finished will lose in price, if left on the hive, from coloring. With those having a less number finished I used to take out those finished and supply their place with sections filled with the extra thin foundation; but of late years the extra work involved in this has made me mostly abandon the plan. Such sections will sell for more mon-ey than they will if left on till the end of the season; but I am not sure that they will sell for enough more to pay for the extra work required in thus taking them. Of course, the whole super can be freed from bees with the escapes, then taken home, and the sections which are filled sorted out, the others being repacked in the supers and taken back to the apiary again; but this makes still more work, and an extra trip to the apiary.

These things are all right where time hangs heavily on one's hands; but with the overworked apiarist, having from three hundred to five hundred colonies in five or six out-apiaries it is better to put all supers not more than two-thirds full back on the hives again. Any super which has been worked in, yet not sufficient to be taken off. Is put back first next the brood-chamber, when a super of foundation-filled sections is set top of it, over which is placed the bee-escape, and the finished super or supers on top of that, so that nearly all of the colonies will have two supers, or

49

eighty-eight one pound sections in which to store from now till the end of the season. If any colony is found which has two supers partly filled, these are both put back and a third super put on, which is empty, except the sections filled with foundation.

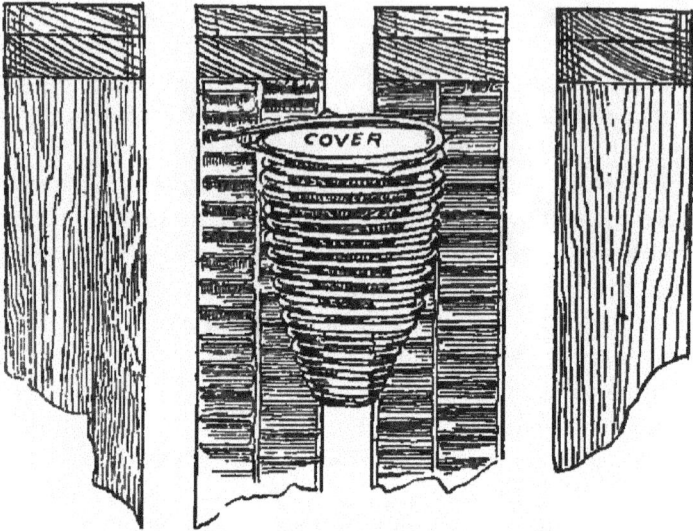

West's Queen-Cell Protector in Use

After a practice of ten years I find that it always pays to keep this empty super of sections on top at all times when there is an expected harvest, as it does no harm except the little labor of setting it on; and as often as one year in three much work will be done in it if it is not filled entirely; and it has much to do with keeping the bees from laying out or being crowded for room, thus doing away with their contracting the swarming fever, as they are quite apt to do when the other supers are nearing completion. Since using this method of keeping an empty super on top I have not had a single swarm during the buckwheat flow, without any further effort at their prevention, while before this I was bothered with nearly half of the colonies contracting the swarming fever during the first week of buck-wheat bloom, they keeping the swarming up till very little section honey would be obtained.

Before going to the apiary at this time I carefully look over the standing of the bee-yard as to the value of the queens in the different hives, as it is given in the little squares on my record-board, and take from the home apiary the number of ripe cells required for use in requeening all colonies having queens which do not come up to the standard of *good* queens. When the sections are all piled on the wheelbarrow, as given above, from a colony having a queen not considered good enough to winter over, I take the opportunity to hunt up the queen and kill her, as she is quite easily found at this time on account of so many of the bees being in the supers just taken off.

Having found the queen and killed her, the next work is to give, them one of the ripe queen-cells I have brought. In taking them from the brooding colony at home, each one was placed in one of the West cell-protectors, so that the bees would not destroy the queen by cutting into the cell before they were aware that their old mother was gone. Bach cell-filled protector was partially imbedded in a sheet of cotton wadding, cut to fit into the bottom of a paste-board thread-box, easily obtained at any dry goods store. Having the number required in the box, another right-sized sheet of wadding is put over all, the cover to the box put on, and a rubber cord sprung around the whole to keep all in a secure position so that the

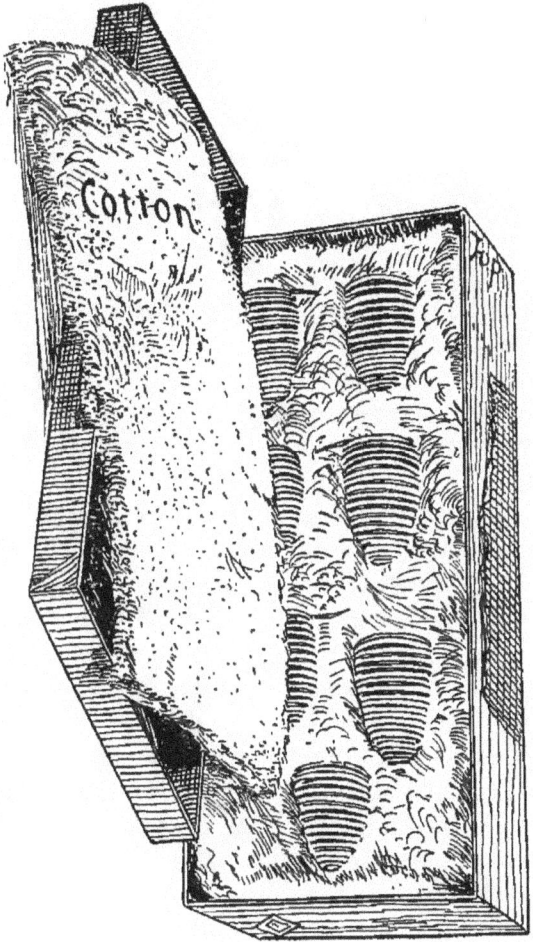

Doolittle's Pocket Queen-Cell Carrier

cells can not roll around when the box is handled. One end of the box is marked *top,* and the base of each cell is placed toward this end of the box so that I may always know that the cells point down when carrying the box in my inside vest pocket, or pocket in my shirt, where cells are always carried at all times except when used in the bee-yard where they are raised.

A "ripe" cell is one from which the queen will emerge in from twenty to thirty hours, and I have often carried such for from one to twelve hours, In the way here given, without the loss or Injury of a single queen. In this work the wadding is far preferable to cotton batting, for the glazing on the wadding keeps the cotton from sticking to the cell or cell-protector, as it is otherwise liable to do.

After killing the queen the frames are all put back in the hive, when two of the center ones are pried apart enough so that the cell-protector will go down just under the top-bar to the frame, when the frames are brought back in place again, this imbedding the protector into the comb so it is securely fastened there until removed by the apiarist. *(See cut on page 50)* As this is the season of the year when the bees do most of their superseding of queens (it seems so natural to them), my loss in using this plan will not average more than one queen-cell out of twenty given. So small a loss will not pay for a special visit to the apiary to ascertain whether colonies so treated obtain laying queens or not — especially as the colony which will occasionally destroy a cell or kill the just-emerged virgin queen have brood of their own from which to rear a queen, so the loss is never very great, should an occasional cell be destroyed. Of course, there is a chance that the young queen may be lost when going out to meet the drone, in which case that colony is doomed unless rescued by the apiarist. In such a case as this the observing apiarist will easily discover this loss by an outside diagnosis of such colonies at a later visit to the apiary. This requeening at this time is so easily done that there is no excuse for having poor queens at the out-apiary.

The reader may think that what is here given conflicts with what I have written in the past about allowing the bees to take care of the superseding of their queens themselves. With the small and contracted brood-chamber, I still hold that the bees will take care of that matter fully as well as the apiarist can; but with this system of working, and that with ten-frame Langstroth hives, a queen will lay nearly as many eggs in two years as she would under the contraction system in three or four years; so that any queen which is more than two years old is almost sure to be played out; therefore I make it a practice with this plan to supersede all queens which are two years old at this time, and in the way given above. This plan is one of strenuousness all the way through, by which we get a multitude of bees in the field at all times during the honey harvests; and even when ordinary colonies are doing nothing, or securing only a living, these rousing colonies are actually laying up stores. Last May, when the colonies as ordinarily worked were living only from hand to mouth, these big colonies at the out-apiary actually laid up from twenty to thirty pounds of stores in the combs above their brood. And then when other colonies were working a very little or not at all in the section supers, these were completing their first forty-four sections, and well at work in the second super of forty-four above. Such work as this is enough to cause the queen to produce all the eggs in her ovaries in about two years; and as the work of superseding as given above is easily done, I think it well pays to kill any queen when two years old, and give a cell to the colony, unless it is a queen that has proven herself of extra value, when I would keep her to breed from the next year, should she live through.

Having the hives all ready for the buckwheat harvest, the poor queen matter disposed of, and the completed supers on the escape-boards, I next attend to any and all the minor things about the apiary that need attention, when the honey is loaded and a start for home is made. If there is more honey than can be carried at one load, it is left right on the hives over the escape-board till I can conveniently come after it; for it is just as safe there as anywhere it can be left, unless we have a building at the apiary for the purpose of keeping honey, which I do not, nor do I consider it needful. If I feared the work of thieves, I would take this honey to the farmer's house, or go back immediately for it; but as it is, I often leave it over the escape-boards for a few days or a week, till some convenient time comes to bring it home.

In the above I have given the reader the work done during the seventh visit to this apiary.

Chapter Eight - Progress in the Supers

Nearly a month has passed since my last visit to the out-apiary, and it is now August 18. The buckwheat is now in full bloom, and the snow-white fields, nestled down here and there among the meadows, corn-fields, and pasture lands, remind one of the days in early spring when the snowbanks are loath to leave under the enlivening influences of the on-coming summer sun. With the blooming of buckwheat, cool days and colder nights come on, which are not what is needed for a good yield of honey from that source. Hot days, with heavy dews, and an occasional foggy morning, are the ideal for a prolific yield of nectar from buckwheat. But the bee-keeper always looks on the hopeful side, seeing the silver lining to the cloud, even though this lining may be on the side from him, and hidden from his outstretched arms. In just such a hopeful mood I am again at the out-apiary, this time to see that all colonies have sufficient room, should there be a heavy flow from buckwheat through returning good weather.

Notwithstanding the poor weather, I find that most of the colonies are well along in the super next to the brood-chamber, while the most of them are beginning work on the foundation in the one above Four or five are quite well advanced in these, and with such the supers are exchanged, the one being nearly completed set over that having less work done in it, with a super of empty sections on top of the two In this case this top super was of no value, as the season was so poor that the bees did no work in It. However, in this race for honey we can not tell how things are going to turn out, and I hold to the idea that it is always better to do a little work for naught than to have a loss of ten to twenty-five pounds of honey from each colony through any inattention of mine. Forty minutes to an hour sufficed for all that was necessary to be done at this time, and the whole gave me an excuse for an enjoyable outing with the auto. This was visit No. 8. If greatly pressed for time, this visit could be dispensed with without experiencing any great loss in honey in the average year.

Chapter Nine - A Simple Way to Put On Escapes without Lifting

It is now September 8th, and the honey season for 1905 is ended, as no surplus is ever secured in this locality from fall flowers. And it has been one of the most singular seasons I have ever known as to poor weather at the time of the blossoming of our honey-producing flora. It was mostly wet, cool, or very windy, during the time of clover, basswood, and buckwheat bloom, our three resources for surplus honey, and quite generally fine and warm outside the time they were in bloom. We often have poor bee weather during the time one of these sources for honey is in

Use of the Wedge between Super and Excluder Board

bloom, and once or twice I have known it thus during two of the sources of supply; but to have it poor during all three puts the season of 1905 at the top, along the line of bad weather, during the expected harvests from all sources, and giving it the name of the "poorest season ever known" among my bee-keeping neighbors. Enough thin nectar was gathered to keep their bees rearing an abundance of brood, resulting in much swarming, and hives light in stores tor winter; but the surplus crop with them was very meager.

I now go to the out-apiary for the ninth visit, and the chief work at this time is to put an escape-board between the brood-chamber and the supers of the whole twenty-eight colonies. To do this best, one of the escape-boards is placed by the side of each hive, before I commence. when I take the piece of wagon-spring used to pull the staples out at the first visit (a long stout chisel will answer in place of the spring), the smoker and a wooden wedge, 1½ inches wide by one foot long, the same being two inches thick at the big end, and go to hive No. 1, row 1, stepping to the back side of the same. The point of the wagon-spring is now pushed between the supers and the hive, or between the supers and the queen-excluder, where one of these has been left on, as with the tiered-up hives. I now bear down on the "handle" end of the spring, enough so a crack is made of sufficient size to insert the point of the wedge, pushing the wedge until a one-eighth-inch opening all across the back is made, when puffs of smoke are driven through this crack to drive the bees away. I am careful not to make this crack big enough at first to let out any bees; for if I do, they are sure to crawl all about on the back side of the hives and supers, to become a nuisance through my killing them, and their stinging my hands during the rapid handling now required. By smoke driven

through this one-eighth-inch crack, the bees are "stampeded" in all directions away from the place where I am at work, and thus are entirely out of the way. By the use of the piece of wagon-spring as a "pry," the wedge is soon pushed in one-half its length, this giving a one-

How to Put On the Escape Board

inch opening into which I can blow smoke, which is now done quite freely. The smoker is now quickly set down, when one hand grasps the escape-board, and by thrusting the fingers of the other into the opened crack, the supers are lifted up at the back end as high as possible without having them slide off the front of the hive, and the escape-board pushed in as far as it will go toward the front of the hive, when the supers are quickly lowered on to it. The smoker is now quickly grasped again, and a stream of smoke sent in at the opening which this has made at the front of the hive by the escape-board not being quite in place. The chisel end of the spring is now caught under the back end of the bottom super, while the other hand grasps the top (forward end) of the cover, when by bearing down on the spring, so as to make a fulcrum of the escape board, and at the same time pulling with the top hand, the supers art easily and quickly slid in their place, so as to cover nicely the escape board. Quickly go to the front, catch the chisel end of the spring under the escape-board, with the other hand at the back, on top of the cover; bear down on the spring so as to make a fulcrum of the hive below, at the same time pulling with the top hand, when the board with its load of supers is quickly and easily brought completely over the top of the hive. If a sort of rocking motion is given to the piece of wagon-spring when bearing down, it will facilitate matters much, especially where there is a heavy load of supers or hives to go on the escape-board. The heavy end of the wedge takes that to the ground and out of the way, immediately upon the lifting of the super, so neither of the hands is obliged to touch it, thus saving one motion when we are in a hurry to get what is needed done before the bees realize what our interference means. The wedge should be made of some kind of hard wood, and be polished smooth. Otherwise it will "broom" up from the heavy pressure that is brought to bear on it in handling supers or heavy hives, three or four stories high, which are filled with honey. In this way I have put the supers in a whole apiary on the escape-boards without killing scarcely a bee or arousing the anger of a single colony. It has taken some time to tell this in writing; but when the "trick" is once learned, it takes but a moment to do it, and that with an ease which seems like magic, even with three or four filled supers on the hives. This is one of the *easy* "short cuts" I use when taking off supers at the end of the season. An editor of one of the bee papers, after seeing me put on escape-boards in this way, wrote a friend about it in these words: "It was a caution with what speed and dexterity he could manipulate the hives and supers. With his practice and skill he killed very few bees, and he did not irritate them either."

I have dwelt on this because it saves so much of the labor and backache required with the usual ways of clearing the supers of bees when taking off honey, at the end of the season. After the whole are treated in this way I am off for home, as "this is all there is to be done at this visit, this being the ninth in number since we commenced operations in the spring.

Chapter Ten - Taking off the Honey and Storing it at the Out-Yard

From two to four days later, in accord with the weather, I go again, the same making the tenth visit, when the supers are taken off, free from bees. I said, "According to the weather," for the reason that a hot, clear day is not suited for the work we must do at this time, when there is no honey coming in from the fields. Robber bees would drive us home long before we could get the work done. The day desired is a cool cloudy one — one so much so that It will keep the bees in their hives. I do not usually go till afternoon, as by noon it can generally be told what the rest of the day will bring forth.

In taking off the supers, those that have no honey in them are piled up top of each other till they are six to ten high, when a cover is put on each pile, and a 25-lb. stone on the cover, where they are left until wanted for use the next year. As many of those having honey in them as I can carry are packed into the auto or wagon, in accord with which I have with me; and if there is more than I can carry they are piled up, as were the empty supers, seeing that each pile is bee-proof, to wait till I can draw them home. The tiered-up hives are now piled away (using the wheelbarrow as much as possible in all this work), the same as were the supers, those being heavy with honey being piled by themselves, and the light ones in a separate place. These are our reserve combs for next year.

"Sweetening" up the Neighbors

I now take off the escape-boards, put the covers on the hives, and store away the escapes for the next year. Some are deterred from starting out-apiaries by what they consider necessary — an outlay for buildings to store things in; for should they continue only a year or two at any place with an out-yard, such buildings would be almost an entire loss. But I do not find it necessary to have any thing more at the out-apiary than a few extra hives and covers, and often all but two or three of these get into the bees' possession before the season's work Is ended. Smoker fuel, smoker, bee-veil, tools, etc., are stored in these hives; and with the finish up in the fall all are piled away as I have given, where they stand right in the bee-yard all the fall, winter, and spring, till they are needed again, the hives and supers giving all the protection that I find necessary in this locality, and all that is needed in any locality, in my opinion, unless it should be the "wild and wooly" West, where thieves are liable to carry off every thing not under lock and key. And even there a few pounds or sections of honey handed out to those living near the out-apiary will generally win for miles around. No one knows how a few sections of honey given to the half-dozen families living near the out-apiary will sweeten for miles around till they try it. The few receiving these little tokens will be jfour friends; and as those further away are the friends of these few, the good words they say about you will make friends of the whole, so that all will almost fight for you, and if they want some honey they will come to you to purchase it, never thinking of taking it otherwise. But be stingy with the product of your out-apiary, so the few nearest it call you "a louse," then there will be no end to the annoyance you will have, and I guess this will apply in nearly equal terms to the home yard as well.

Weighing Up the Hives

When I come to the colonies which were tiered, I weigh them, as some are liable to be short of stores, through storing too much in the combs above; and any that are light are supplied with plenty by giving them full combs taken from the "heavy" pile in exchange for their light ones. I do not now look after the stores of those that worked in sections, as it is seldom that there is a lack with any of these, as the plan used, together with the ten-frame hive, nearly always causes the storing of plenty of honey for winter. If, when turning the bottom-boards for winter, at our next visit, any are found to be light, a change of heavy combs for some of their light ones is made, so that all are known to have 25 lbs. or more, which is amply sufficient for all their needs till they can be looked after in the spring, when starting them on their road to prosperity, for both themselves and owner. After again carefully looking over all the piles of hives and supers containing honey, to see that there is no crack or hole about any of them sufficiently large for the entrance of a bee, and giving a general glance over the whole, to see that all is in good condition for leaving, I am ready for my journey home. And this is what was done on the tenth visit.

A Retrospect

As I am about to leave I cannot help taking a last lingering look at things, as they have so changed since I came at noon. Instead of tiered-up hives, and those with supers, which have gradually grown up with me during the summer's work, all have assumed the appearance of what they had in spring, and I am reminded that the work of the bees is over till another year. A sort of sadness steals over me, and I fall to wondering if both bees and myself will be alive to work so happily together another year. The merry hum, and the fragrance from the hives, which greeted me when coming to the yard during the summer, greet me no more. I find myself wishing it were spring again, and that I were just commencing the fun of working the out-apiary for another year. I seem to see the bees at work again as they did on those bright "clover and basswood" morns. It seems like a real living picture again — a picture fairer than thought; a picture fairer than a dream; a picture with ten thousand pearls glistening in earth's rarest sunlight, on one stretch of verdure green, and reaching out beyond the winter's vale to the bright spring again, when the butterfly begins to flutter in the pleasant breeze, and the joyous children are chasing after sunbeams. Thus I dream. As I have been musing, the clouds have parted in the low west, and the setting sun has dropped down into the clear space between them and the horizon, throwing over hill and vale ten thousand times ten thousand glittering hues that glow and shine to beautify the landscape and cheer the heart of man. Dawn tiptoes over the mountain tops, and peeps into the valley far below with eager, tender eyes, while darkness gathers up her sable robes to skulk and hide away into the crevices and mountain caves; but in the evening come the long light sunrays, beautiful, to gild the world and gladden it with kisses, lovelier, sweeter far than the rarest, gentlest kiss of dawn. So, too, the evening tide of life may grow more beautiful and blest, if life is rightly lived, believing upon Him who was and is the *light* and *life* of men. And the bees, now in the evening tide of 1905, are enjoying a rest sweeter by far than their restless sleep during the dawn of their activity, six months ago.

59

"Hello there! Gone to sleep?" comes in stentorian tones from my farmer landlord, and I am aroused to the fact that it is fully time that I be on my journey home.

Chapter Eleven

It is now October 10th, and one of those beautiful clear days with enough of smoke and "haze" in the atmosphere to give a balmy air, which makes one of our fall days in New York so delightful. The leaves, which are soon to fall from the trees, all gorgeously arrayed in their many-dyed hues, are made more enchanting to the eyes by being "kissed" by the morning sunshine — surely a splendid day for an auto ride; and, to combine pleasure with profit, Mrs. D. and myself are soon traveling at an easy "pace" toward the out-apiary, breathing the healthful ozone of an autumn day, and feasting our eyes on the ever varying changes of the landscape before us. We go on a roundabout road, instead of the direct one usually traveled, so as to see new scenes; but even this, and with the gait of the auto so slow that I, the driver, need

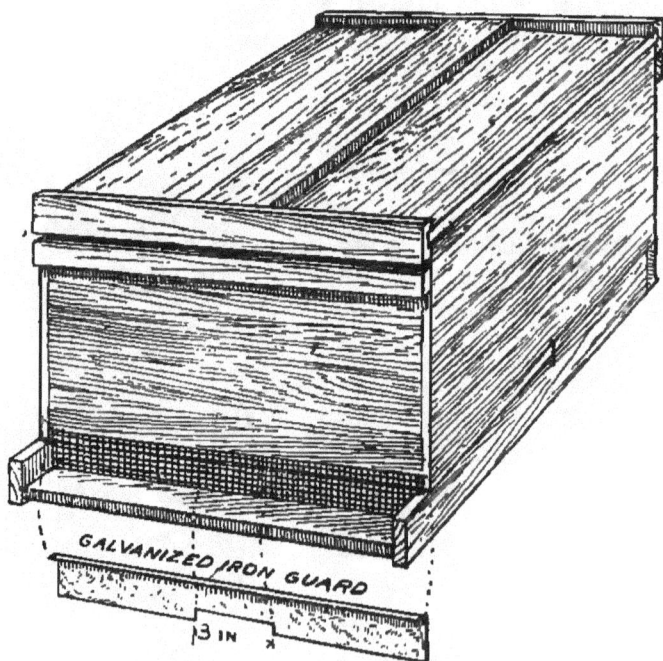

Mouse-proof entrance; 5-8 mesh; bottom-board winter side up; hive-fastener with staples

not be very closely confined to the chauffeur part of the matter, causes us to arrive at our destination all too soon. Mrs. D. goes in to have an agreeable hour with the farmer's wife, while I hie me away to the bee-yard, the most delightful spot in all the world to me except my home, the Sunday-school, and the church of the Lord Jesus Christ.

With a swinging motion of the hands and forearms, together with a sort of backward bend, while the elbows are on the knees, hive No. 1, row 1, is "swung" from its stand to the ground, immediately by the stand's side. A reserve bottom-board is now placed on the stand, winter or deep side up, when a right-sized piece of galvanized wire cloth having a 3/8-inch mesh (the same being used as a mouse-guard) is slipped into the saw-kerf made for it on the inside edge of the two-inch strips, which holds the hive that far from the board below. A few puffs of smoke are now blown in at the entrance of the hive, when the point of the ever useful piece of wagon-spring is thrust into the same, and, with a lifting motion, the bottom-board is made to part from its place through the breaking of the propolis which has been used during the summer to fasten it there.

With the same swinging motion, as before, the hive is almost instantly on the newly prepared bottom-board, and brought forward till it touches the mouse-gard of 5/8-inch-mesh wire cloth. When the bees are wintered at the farmer's cellar, who owns the land the

The Lifting-Swinging Motion

out-apiary is located on (and I should always winter them there if possible), this mouse-guard is an absolute necessity, as a former experience of rat-and-mouse-destroyed combs and bees told me. Hive No. 1 now has an entrance two inches deep the whole width of the hive, all open except the wire cloth. This must be tightly closed in some way for a month or so, or until the bees are set in the cellar, to prevent robber bees from gaining access to the honey in the hive. This is best done with a piece of galvanized iron, the same size as the mouse-guard, having a piece three inches long by ½ inch deep cut from the bottom side of it, when it is slipped down in the saw-kerf on the outside of the guard.

Having No. 1 thus ready for cellar wintering, the bees on the bottom-board, if any still adhere, are jarred off in front of the hive, and I go to No. 2, treating it in the same way I did No. 1, only using the bottom-board from No. 1 instead of a "reserve" by turning it deep side up. In this way I keep on till all are thus treated.

By this swinging process, as here given, which I always use in changing the bottom-board both in fall and spring, there is not half the fatigue and none of the backache that are experienced by the usual way of lifting hives which are heavy with honey; and I would recommend it to any and all, in any and every place where it can be used.

In this change of hives and bottom-boards, any that are light in stores are quickly detected; and if any such are found, they are so marked as I go along. I do not find any of these light colonies oftener than once in three or four years; and when I do, all that is necessary is to open the hive and take out one, two, three, or four of their nearest empty combs, and give them as many heavy ones from the reserve pile in giving heavy combs of sealed honey at this time of the year I think it better to alternate them with the light ones which the colony has, where more than one are given, as I consider such alteration more in harmony with good wintering. Of late I have been trying a little different plan where colonies are light in stores, which is, to set the heavy combs of honey next to one side of the hive, but having just one light comb next to the wall of the hive. Suppose I am to set in three heavy combs. I first take out three that are the nearest empty, shaking the bees from them. I now draw one of the light frames next to the side of the hive, when the three heavy frames of honey are put in. This brings the outside of the cluster in touch with the first heavy comb of honey; and as soon as they are in need of more food than that contained in their light combs they begin to move over on the heavy ones. Thus the cluster moves toward their stores all winter, and never starve. With stores equally divided on either side of the cluster, that being in the center, it often happens that the cluster moves toward one side; and when the honey on that side is consumed they fail to cross over to the opposite side, and so starve with plenty of honey in the hive, but seemingly out of their reach.

Having the bottom-boards all turned, and knowing that all colonies have plenty of stores, I next fasten all the bottom-boards to the hive by driving a crate-staple on either side, with one point going into the hive and the other into the bottom-board, as near the center as is convenient with rapid working. Some seem to think that it is better to use one of these staples at each corner, and this may be so where hives are to be hauled from the out-apiary home for wintering, and back again in the spring; but for carrying to the cellar, and setting out again, the carrying being done by two men and a rope, the two crate-staples are amply sufficient. In driving these staples I find that a hand-ax, or something having a driving-face sufficiently large to cover the whole staple at each blow, thus driving both points at the same time, is much better than an ordinary hammer that drives only one point at each blow.

With the driving of the last staple the work done at the eleventh visit is accomplished, as well as that for the year, except setting the bees in the cellar. As the day still continues fine, we take another roundabout road for our ride home, where we arrive in due time, feeling that the day has been very profitably spent, even though we have consumed the most of it on the road.

Chapter Twelve - Closing Words; Further Suggestions to the Plans Given in the Preceding Chapters

After using what has been given in the previous visits, both in the home yard and out-apiary (for the plan is equally good for the home apiary) In its different stages of growth, as it developed during the ten years between 1889 and 1900, and pretty much entirely for the past five years, I wish to say that I believe it ahead of any and all other plans used up to the present time, in that it gives the largest possible number of bees at the right time for the harvest, with little or no disposition to swarm; controls swarming perfectly, puts all honey not needed for the rearing of bees or winter stores in the sections, and that with the least possible work that can be used when working for section honey. Doing this it is of great value in the home apiary, and an actual necessity for an out-apiary worked for section honey. An additional value that attaches itself to the plan is that the sealing or cappings of the honey in the sections are nearly or quite as white as those where honey is built by new swarms where they are hived in contracted brood-chambers having only frames with starters in them below, which all know is of a whiteness heretofore secured in no other way. This fact alone would be of sufficient value to pay any bee-keeper for adopting it, even if it were not "head and shoulders" above any thing else in securing a big crop of section honey without any swarming.

The cause for this white capping, as I view it, comes from the bees fully cleaning, perfecting, and partly or wholly filling the combs along their tops, with honey, which, later on, after the "shook" swarming has taken place, become their brood-nest; or when these combs are occupied for their brood-nest proper, none of this cleaning of old combs is indulged in, or cappings from over emerging young bees handled, to carry bits of old comb or travel stain into the sections while they are being capped, as is the case with all other ways of using old combs. I have noticed for years that, when bees are cleaning old combs, or where much brood is emerging near the top-bars to the frames just under the sections, more or less of this refuse matter is worked into the cappings to our section honey. Even where new swarms, hived on starters, put brood next to the top-bars to the frames under the sections, the cappings to such sections as are sealed after this brood begins to emerge are not nearly so white as it was previous to this— especially along the comb in the sections near the bottom.

Then the labor part in producing section honey by the plan as here outlined is much less than with any of the other plans recommended m our bee books and papers, so far as I have tried them, and I have tried nearly all. A man of usual working ability should be able to work five out-apiaries, in connection with the one at home, with little if any help except, perhaps, a few days when he is making swarms and setting the bees in and out of the cellars. Were I from 25 to 40 years old, and free from the rheumatism which I now enjoy (?), I should not hesitate to undertake the working of six yards containing from 50 to 75 colonies each, including the home yard. But my age, and rheumatism in back and knees, to an extent which makes it very difficult to "get to going" every morning, and often with only pain and weariness during the whole day, prohibit me from taking a very active part in these matters much longer.

After preparing, crating, and marketing the honey produced by the sixteen colonies at the out-apiary, worked as has been given on the preceding pages, I summed up the product and found it as follows:

Section honey sold..............................1763 pounds.
Given to neighbors...................................42 pounds.
Kept for home use...................................27 pounds.
Total..**1832**

This divided by 16, the number of colonies worked for section honey, gives the average product of each colony as 114½ pounds, and that in a season when my bee-keeping neighbors report but very indifferent success. Had the season been good during the bloom of only one of the honey-producing flora this could easily have been 150 to 175 pounds, while good honey weather during all of the bloom would doubtless have chronicled an average of 250 to 300 pounds. There is also about 500 pounds stored in the reserve combs, ready for turning into bees, etc., next spring, which is fully as much as was on hand a year ago, besides an increase of nine good colonies.

Through sickness at "shook-swarming" time, as given elsewhere, and the generally poor season, the yield was only 105½ pounds of section honey per colony on an average in 1906. And here Doolittle must take a back seat already and that with his own plan, for a report has come to me from a party working with the plan (through reading the serial in *Glean-*

ings in Bee Culture) of a yield of 135 lbs. from each colony on an average; while another reports an average of *three times as much* from the colonies worked by this plan as from those worked on *his* most approved plans of the past.

As I see it, this yield of 114½ pounds per colony in a poor season came from three reasons. First, the great number of bees in each hive at the commencement of the harvest. A careful estimate of the emerging bees in hives worked on this plan in the home yard, where I could more certainly verify these things by opening a hive or two set apart for such work, every day or oftener, if I thought it necessary, would give 76,431 bees on the stage of action at the time of the first "shook-swarming," barring accidents. Then should we allow 16,431 for these accidental deaths, which would be a greater loss than I would think possible, we would still have 60,000 bees as the number to commence work in the harvest from white clover, which is a mighty army, sure.

The second reason was, that these 60,000 bees had no desire to swarm, so they worked with great energy on every and all occasions, when there was a day or hour, even, when it was suitable for a bee to go out, or for the secretion of nectar.

The third reason was, the giving of super room enough, and in such a way that it encouraged them to the greatest activity, kept them from contracting the swarming fever, and at the same time did not, at any point, discourage them from entering this room nor cause them to retreat from any of the room which they had commenced to work in. This giving of storage room, in a way advantageous for the best work, either in a light or heavy flow of nectar, both before and after our "shook" swarming is done, is an item which has not sufficiently entered into the plans of the past. By this plan the bees are at work in a second hive of combs before they hardly know it; and at the time of our making them swarm, the whole of that "mighty host" are ready to take immediate possession of the sections through their previous occupation of a "super hive," which *now* becomes their "richly endowed" home or brood-nest.

The plan of coaxing bees to an early work in the sections, and at the same time retarding swarming by giving an extracting-super for a short period before the opening of the honey harvest, and on the advent of such harvest taking off this super and putting on the sections (this causing the bees to enter readily the sections from having previously worked in the extracting-super) was originated some years ago. But that plan did not place the honey stored in this extracting-super in the sections, nor prevent the swarming of the colonies so treated, later on; but generally right in the height of the honey harvest, when swarming is the most injurious to the prospect of a crop of section honey, hence was only a step in advance of the older ways of working.

To emphasize a little: The beauty of the plan I have now given is, it puts in the sections *all* honey not *actually used* by the bees, and that with no swarming during the honey harvest, or previous thereto, and that with the *largest possible* force of bees in a colony, consistent with working for section honey.

And I wish to speak a little more about using "bait" sections in the first super put on at time of "shook" swarming. I prefer to use at least twelve of these, so that the bees will immediately enter the sections with their loads of honey that the queen will oblige them to remove from the combs they are shaken on, so that she may have room for her eggs. *Failing here, at the start, would cause a failure in the whole,* for this honey *must* be removed if the queen is to keep right on with her prolificness; and by having a place for the immediate storing of this honey In the first super above the honey-filled brood-nest, the bees not only carry this honey out of the queen's way, but they also carry that gathered from the field to the sections, this causing the immediate drawing-out of the foundation in the sections, other than the baits, so that there is a start made all along the road toward success within one hour after the bees are shaken from their brood. Here is one of the *great* advantages of this plan, and one of the things *original* with it. And I did not realize till one year later, 1906, that, if the bees contracted the swarming fever before they were shaken sufficiently to cause the queen to slacken her laying, that it, in a measure, detracted from the plan by her not filling every empty cell with eggs as fast as the

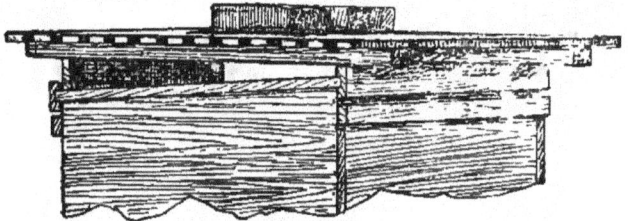

A Doolittle Shade-Board

bees removed the honey from them to the sections; for till sickness prevented me from doing the "shook swarming" in time I had had no such slackening, as no colonies had advanced far enough for the queen to stop laying in preparation for her flying with the swarm. Therefore, I consider all of the advice given in the past, "to wait about shaking till preparations for swarming are made," as decidedly wrong. But the upper hive of combs keeps the desire to swarm down till it is time to shake for the clover harvest, so there is no need of a failure here if we do our "shook swarming" when it ought to be done.

Another thing, which I see I failed to mention in any of the accounts given of the different visits, which I consider a great help in any apiary, is shade-boards. I am convinced that a colony of bees will do much better work where the hive stands right out in the sun during the whole season, except as it is shielded during the middle of the day by a shade-board. I make this board of half-inch lumber, 20 inches long, nailed to two strips 7/8 thick by 1½ wide by 28 inches long, covering the whole with a sheet of 20x28 tin. Roofing-paper will answer nearly as well as the tin, if kept painted. Near one end of the shadeboard, and before putting on the tin, I nail, on the under side, a piece of 7/8 stuff 6 inches wide by 20 long, nailing down through the board into the edge of this twenty-inch piece. When the board is on the hive, this last-named piece rests, by its lower edge, on the back part to the cover to the hive, while the cleats rest on the front part to the cover. This gives this shade-board a "pitch" towards the front, or south side of the hive, so it will carry off all rain, shade the hive mostly from 10 A.M. to 2 P.M. each day, and allows the air to circulate freely all over and about the top of the hive, so that the bees are never driven out of the sections through extreme heat, as is often the case where hives stand in the sun without any shade, even though the cover is painted white. It matters little what color these shade-boards are painted, on account of the circulation of air under them; still, where I paint hives at all I prefer the color to be white.

I feel that I ought not to close this work without saying a few words regarding the automobile for the apiarist, inasmuch as I have mentioned it several times when telling of my visits to the out-apiary. At times I think the one I have (an eight-horse-power single-cylinder Pierce Stanhope, and I think it is as good as any, or I would not have purchased it), an expensive luxury. At other times I consider it the nicest thing in the world to travel in, both for pleasure and profit; and at other times I consider it almost a necessity for me in my apicultural work. The time when I consider it an expensive luxury is when the roads are in a condition not suited for its use, on account of deep mud and snow, which is fully six months in the year in this locality. If I lived in a city or a country where the travel on the roads did not cut them up so the mud is from three inches to a foot deep, or where the snow did not fall or drift from one to ten feet deep, this "expensive luxury" would not play such an important part. Then the auto could be used nearly if not quite all the time, thus saving the keeping of a horse, wagons, and sleighs. But as it is, I must keep these in addition to the auto, at an expense of from $200 to $300 a year. The times when I consider it both pleasurable and profitable is when the roads are good, enabling me to go to and from the out-apiary, and elsewhere. In less than half the time consumed by the horse, with no trouble from flies tormenting, bees stinging, or fright from any strange thing along the road, as is very often the case where a horse is used as a means of conveyance. And I can carry as many supplies to the apiary, or bring home as much honey with it, at a trip, as I can when using the horse.

In all of these hauling operations, blankets are used to keep from marring and injuring the auto. The times when I consider it a necessity is when I wish to drive right up to or into the apiary for loading or unloading stuff; when I am short of time, and must get to any place very quickly, and when I wish power for running machinery, although I have not so much need of this as formerly. At no time do I appreciate the auto more thoroughly than when I drive it right into the bee-yard for loading and unloading heavy stuff. The auto is low-down, so I do not have to lift things so high in loading as with a wagon. Then with the horse I must do a lot of lugging and carrying myself, or else get some one to help me draw the heavily loaded wagon to a safe distance from the bees, and, even at the best, have a constant care not to get the horse stung. Nothing of the kind with the auto, for I have never Known of a bee stinging it.

Then if the washing-machine, grind-stone, churn, feed-cutter, emery-wheel, planer, buzz-saw, etc., are to be used, just back the auto up to the proper place, "jack up" one of the hind or driving wheels, and "block" the other, so as to make the auto stationary, connecting by belt the jacked-up wheel and the machine you wish turned, when everything is ready for "the start." Does the machine need to be run slow? Set the spark-lever at *slow*, or at the place you put it for slow running when on the road. Do you wish a 3,000-a-minute gait for your buzz saw or planer? Set the spark-lever at a 25-mile-an-hour gait and you are at home in the matter.

The main thing in running machinery with the auto is to know how to time the matter of oiling the latter. The instruction-book which comes with the auto will tell us how many miles it is to be run to once oiling of certain parts. But the auto is not making "miles" now, but revolutions. By getting the number of inches the revolving wheel is in circumference, and dividing the number of inches there are in a mile by it, we can ascertain the number of revolutions the wheel would make in going a mile on the road. Then by multiplying this by the number of miles we were allowed to run for once oiling, we shall have the number of revolutions the jacked-up wheel can make without danger from lack of oil. Now with a speeder find the number of revolutions the drive-wheel is making per minute when the spark-lever is set for different degrees of speed, when it is easily told how many minutes or hours can be safely run on one charge of oil or grease. Where any apiarist lives in a location where the auto can be run the most of or all of the year, he can well afford to let his horse and wagons go, and purchase an auto; but if in a locality like mine, then it is best to ask, "Can I and my family afford both?" before buying one.

www.ingramcontent.com/pod-product-compliance
Lightning Source LLC
Chambersburg PA
CBHW032017190326
41520CB00007B/508